中国建筑彩画图集

修订版

主　编　何俊寿

副主编　王仲杰

编辑委员

韩振平　马炳坚　蒋广全

杜恒昌　赵双成　高成良　冯世怀

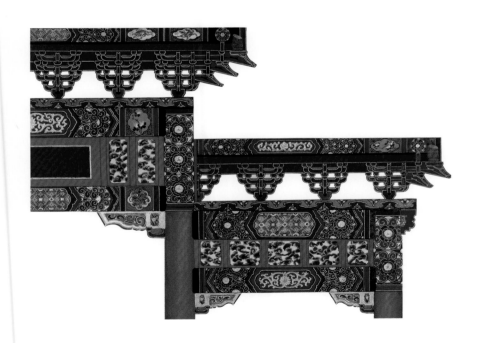

图书在版编目（CIP）数据

中国建筑彩画图集 / 何俊寿主编. —天津：天津大学出版社，1999.4（2014.1 重印）
ISBN 978-7-5618-1165-8

Ⅰ.中… Ⅱ.何… Ⅲ.建筑艺术－绘画－中国－图集 Ⅳ.TU204

中国版本图书馆 CIP 数据核字(1999)第 11648 号

中国建筑彩画图集（修订版）

主　　编：何俊寿
副 主 编：王仲杰
编辑委员：韩振平　马炳坚　蒋广全
　　　　　杜恒昌　赵双成　高成良　冯世怀

出　　版：	天津大学出版社
出 版 人：	杨欢
地　　址：	天津市卫津路 92 号（天津大学内）
邮　　编：	300072
制　　版：	北京精制轩彩色制版有限公司
印　　刷：	北京信彩瑞禾印刷厂
经　　销：	全国各地新华书店
开　　本：	230mm × 302mm
印　　张：	29
字　　数：	341 千字
版　　次：	1999 年 4 月第 1 版　2006 年 1 月第 2 版
印　　次：	2014 年 1 月第 4 次
印　　数：	11 601－13 600
定　　价：	288.00 元

如有印装质量问题，请与本社发行部门联系调换。

正、副主编及作者

赵双成　高成良　蒋广全　王仲杰　何俊寿　马炳坚　杜恒昌　冯世怀

编辑委员在审评入选作品

序

我国古代以木构建筑为主，为了保护木材免受雨淋日晒，很早以前就知道用油漆涂刷房屋。从战国时期开始，封建帝王的宫殿建筑无不涂漆画彩，以示华丽，可见中国古建筑最讲究色彩美丽。西汉文学家张衡在《西京赋》中描写当时皇家宫殿建筑"屋不呈材，墙不露形"，意思就是说，木构和墙体表面不可暴露原材，必须施以油彩刷饰才符合使用要求。

文献记载和考古发现都证明建筑彩画在我国有悠久的历史传统，它与雕塑、壁画同属姊妹艺术，与建筑有十分密切的关系。一座房屋的形体美要与色彩美同时并存，才能使人产生美感。从一定意义上来讲，建筑彩画也是一种美化生活环境的文化载体。因此，自古以来"雕梁画栋"便成了形容建筑美丽的代名词。

过去，这种实用艺术一直掌握在民间艺人手里，被称作"匠画"，它与"文人画"有显著差别。在悠久的历史长河中，民间的艺术家们从实践中积累了丰富的创作方法和技艺方面的心得经验，涌现了不少杰出画手。可惜那时彩画艺人因为文化程度所限，虽身怀绝技，却未能留下完整的文字著作，十分遗憾。此外，这行手艺一向采用师傅带徒弟的方式来培养人才，师徒之间口传心授，边学边做，从来不著文字，技术交流范围有很大的局限性。

但长期历史积淀，却给后人留下大量的艺术语言——"画诀"（即口诀）。学徒出师之后，掌握了这些绘画语言，并熟记工艺要领，就能独立进行创作，成为行家里手。还有画店、作坊里的老艺人也曾留下为数不多的样稿（即"粉本"）。所有这些流传在民间的画诀或粉本都是十分珍贵的文化遗产，我们应该及时进行挖掘整理，贡献给社会，以防人亡艺绝，流于失传。

彩画这门工艺美术发展到清代中晚期，就北京地区而言，以和玺五墨彩画、旋子彩画和苏式彩画最为盛行，在"画作"中几乎占据了统治地位。至清代末期，更盛行苏式包袱彩画，它的图案内容丰富多彩，凡楼台、山水、人物、翎毛、花卉、博古及传奇故事皆可入画。它以其构图轻松活泼，色彩绚丽，颇受人们的青睐。因此，从那时起，这种建筑彩画在园林、宾馆、商店和高级住宅中极为盛行，可谓蔚然成风。

本图集共搜集到传统彩画样稿及实物照片计150余幅，在当前建筑装饰的发展中，为彩画界同行进行创作提供借鉴与参考。同时，还请彩画专家王仲杰先生为本图集撰写了论文一篇，详细阐述了明、清彩画的工艺特征，希望能使广大读者在理论方面有所收益。这本图集倘能对您有所帮助，我们将感到十分高兴，故此乐为作序。

国家文物局高级工程师　杜仙洲
1999年1月于北京安贞里

Preface

Chinese ancient buildings were predominantly wooden structures. In order to make the wood weatherproof, ancient Chinese people learnt to paint their houses from remote antiauity. As late as the Warring States period (475-221 B.C.) palaccs were all painted and drawn to show their magnificence. This evidences that ancient buildings were particular abot bright and gay colors. In his descriptive prose "Xi Jing"(West Capital), Zhang Heng, the scholar and seientist of the West Han Dynasty (206-25 B. C.) wrote,"The house does not expose its materials, and the wall does not expose its original form."It means that the surface of the material of the wooden structure and the wall body must not be exposed and must be drawn and painted to satisfy the requirements of use.

Written reeords and archaeological finds show that architectual decorative polychrome paintings have a long history. They are closely associated with buildings and are one kind of the sister arts like sculpture and freseo. Only the co-existence of the beauty of form and beauty of colors imparts us the sense of beauty. In a sense the decorative polychrome paintings is one of the carriers of culture to beautify the living environment. Therefore, the phrase"carved beams and painted rafters"has become the synonym of magnificent masions since old times.

In the past this practical art mastered by the folk craftsmen was called "craftsman's drawing", quite different from the "scholar's drawing". Through centuries the craftsmen had accumulated a world of experience in design methods and techniques in practice and hosts of outstanding painters sprang up. The pity is that, being poorly educated and even illiterate, they could not put down their experiences and skills in written forms though they might possess unique skills. Such skills had always been taught by master workers to their apprentices. Apprentices learnt skills through doing under the oral instructions of their master workers. Technical exchange was very limited.

However, the age-old oral instructions had gradually come to be a kind of technical language-rhymed formulas for drawing. After serving his apprenticeship, learning the formulas and bearing them in mind, the apprentice was in a position to undertake the creative work by himself. Moreover, veteran workers in drawing stores and workshops handed down a small number of master drawings. The formulas and master drawings are precious legacies for which we should search and sort out in time to serve the society lest they could be lost.

As a kind of industrial arts, three types (Hexi Pattern, Xuanzi Pattern and Suzhoustyle Pattern) were prevalent in Beijing area in the middle and late periods of the Qing Dynasty. By the end of the Qing Dynasty the Suzhoustyle Pattern was all the rage. Its patterns were rich and colorful, including landscape, figures, birds, flowers, curios and even legendary stories. They were well received for their lively composition and gorgeous colors. For this reason such decorative polychrome paintings have a vogue in landscape gardens, guesthouses, stores and luxury residences.

This collection contains over one hundred master drawings of traditional decorative polychrome paintings and photos. They will be of value in the development of the decorative polychrome paintings and workers in this trade will benefit from them.

Furthermore, Mr. Wang Zhongjie, a specialist in the decorative polychrome paintings was invited to write an introduction to this art, in which he gives a detailed description of the characteristics of the decorative polychrome paintings in the Dynasties of Ming and Qing in the hope that the introduction will help readers understand this art in theory. We shall be delighted if this collection can be of value to readers.

Du Xianzhou

Senior Engineer

National Bureau of Cultural Relics

January 1999

Beijing

前　言

改革开放20年来，中国古建筑事业得到了复苏和发展。运用中国传统建筑技艺，维修保护文物建筑和人类文化遗产、修缮复原名胜古迹、完善繁荣历史文化名城等工程项目日益增多，建造中式建筑和具有传统风格的新建筑在各城市建设中亦占有一定的比例，从而促使中国建筑彩画专业成为学习、研究、运用不可缺少的技艺项目和工作。由于在保护维修文物古建筑中要严格执行《文物法》的规定，保证做到不改变原状，因而必须遵循"法式"，使建筑的彩画制度、品类及图式与之完全符合，工艺、材料亦不容改变。而仿古建筑也需要绘以传统彩画来体现不同历史时代的建筑格调，使运用合宜。新形式建筑在梁、枋、柱、墙边、藻井、天花等室内外装修上以中国建筑彩画为蓝本，加以改进，设计、绘制与其部位、功能一致的新式彩画，如运用得当则会起到"画龙点睛"的理想效果。

基于这种实际需要，以《古建园林技术》期刊作者为主的建筑彩画名师们，在近20年间临摹、复原、整理再现了许多中国建筑彩画，精绘了部分典型的建筑彩画"样稿"，并创作了一批新式彩画。

此次，邀请八位专业人员作为编辑委员，精选汇编了这本《中国建筑彩画图集》，以奉献给文物保护、建筑、园林设计与施工、装饰装修以及工艺美术各界的朋友，供参考酌用。

本图集按建筑彩画年代，分类选编了古建筑彩画"样稿"作品100幅、彩画实物照片53幅，介绍了中国建筑各部位构件的彩画种类、式样及组合关系。此外还择录了各类墨线图100余幅，便于读者对中国建筑彩画从设计底图到绘制完成有较系统、全面的了解，从而使图集具有学习参考、运用研究、欣赏收藏等多种作用与价值。

本图集内各幅图片、照片均标注题目，且在各类"样稿"作品之前由绘制编纂人写有一篇精练的短文，作为一类彩画的介绍，而不再逐幅说明，以避免重复。

图集还选入6页"地方彩画"。篇幅不多，意图在于表明在明、清时期除以北京地区为代表的官式建筑彩画之外，我国各地尚有承袭原有特色的、种类繁多的地方建筑彩画。这些作品是中国建筑彩画百花园中的群秀之一。

新式彩画是20世纪中西建筑相互融合，在中国新建筑中兴起的一类建筑彩画品种，50年代以后得到广泛应用，预计今后仍会有所发展。本图集选辑了19幅"样稿"作品，以充实建筑彩画门类，也为"引玉"之举。

图集中实物照片及墨线图均为编辑委员拍摄与绘制，彩绘"样稿"大部分亦出自这些成员之手。此外，尚辑录以下作品：
（1）前"文物整理委员会"所刊印《中国建筑彩画图集》之册页；
（2）已故老画师郑书本先生等遗作；
（3）北京老画师郭汉图老先生之佳作；
（4）北京彩画名师边精一先生之佳作；
（5）北京市园林古建工程公司王忠福、张京春等先生绘制的"样稿"；
（6）山西省太原市园林古建筑工程公司岳俊德、白永红等先生的作品。

特在此言明，并向有关单位及各位先生表示敬谢之意！

在本图集辑选与编写过程中，蒙北京市古代建筑工程公司暨所属油饰彩画分公司、《古建园林技术》编辑部等单位及工作人员的鼎力支持和协助，在此致以衷心的感谢！

古建筑、文物老专家罗哲文先生为本图集题写书名，杜仙洲先生为本图集作"序"，并对这项工作加以鼓励和肯定，在此谨致谢意！

由于绘制撰写成员水平所限，本图集如有疏漏不当之处，敬请读者教正。

何俊寿　撰
1999年1月

Foreword

In the past 20 years of reform and opening-up the construction business of ancient buildings has revived and developed. Ever increasing are the engineering projects of repairing and protecting cultural buildings and human cultural heritages, renovating and restoring scenic spots and historical sites and improving noted historical cultural cities by means of traditional Chinese building technology.Constructing buildings modelled after antiques or new buildings with traditional style accounts for a certain proportion in the urban construction of cities in many places.Such situations spur the trade of the decorative polychrome paintings in Chinese architecture to be the essential technological subject and work in the in heritance,study and application of decorative polychrome paintigs.The stipulations in " The Cultural Relics Law" must be strictly put to operation in the course of protecting and maintaining ancient buildings to ensure that the original form of the ancient buildings will not be changed and therefore the"standards" must be observed so that the systems,types,and patterns of the decorative polychrome paintings can be in complete accordance with the ancient ones,and the process and materials cannot be varied.The colored ones must embody the architectural styles of different historical periods and must be appropriately copied.New decorative polychrome paintings to be painted on beams,rafters,columns,ceiling,ceiling central cavity and wall edges of new buildings must be based on the traditional drawings and must be modified and redesigned to suit the position and function of the structural components in the new building.Proper handling of the decorative polychrome paintings will bring effect of "getting twice the result with half the effort."

To meet practical needs the editor of the journal Technology of Ancient Building and Landscape Garden and noted painters of colored drawings have copied,restoreds,sorted out and produced large numbers of traditional architectural decorative polychrome paintings in the past 20 years.They have also drawn elaborately part of the typical "master drawings" and created new decorative polychrome paintings.

This time 8 professionals are engaged as editors to select and compild this collection of Chinese architeetural decorative polychrome paintings for departments in charge of the protection of cultural relics,construction and design of landscape gardens,decoration and fitting-up and industrial arts.

In the collection 100 master drawings and 53 photos of the real decorative polychrome paintings in the ancient buildings are arranged according to their date of drawing and classification,which show the types,patterns and the combined relation of this colored drawings painted on components in different positions.Besides,more than100 pen-and-ink drawings of different types are provided for readers to have a systematic and all-round knowledge of the whole process from designing the draft to the completed work.This collection can be helpful to those who are interested in learning,studying,appreciating, applying and collecting the decorative polychrome paintings.

Each drawing and photo is captioned and preceding each type of the master drawing is a brief description.The caption of each drawing is omitted.

Added to the above-mentioned drawings are the 6-page "Local decorative polychrome paintings", which,though meager in number,epitomizes diverse, characteristic decorative polychrome paintings flourishing in places other than Beijing area where official paintings were dominant in the Ming and Qing Dynasties,The local decorative polychrome paintings are also beautiful flowers in the garden of hundred flwers in their own right.

The new-type decorative polychrome painting is a new genre of the decorative polychrome painting and the result of the fusion of the Chinese architecture and western architecture in the 20th century.It has been widely used since1950s and will develop in the future.Nineteen master drawings of this type are included in this collection as the first batch to attract more works.

All the photos and pen-and-ink drawings were taken and drawn by our editors. Most of the master drawings were drawn by them, too.

In addition, included in this collection are the following drawings:

1. Works taken from "Collection of Architectural Decorative Polychrome Paintings" published by the Sorting-Out Committee of Cultural Relics.

2. Works by late painter Mr. Zheng Shuben.

3. Works by Mr. Guo Hantu.

4. Works by Mr. Bian Jingyi.

5. Master drawings by Mr. Wang Zhongfu, Mr. Zhang Jingchun of Bei jing Ancient Building Construction Engineering Co.

6. Works by Mr. Yue Junde, Mr. Bai Yonghong, et al. of Construction Engineering Co. of Ancient Building and Landscape Garden, Taiyuan, Shanxi Province.

We are indebted to the above-named painters for permission to reproduce their works.

I would like to express my thanks to the following institutions and personnel concerned for their support and help in selecting and compiling this collection :Bei jing Ancient Building Engineering Co. and its Painting and Decorative Polychrome Paintings Branch Co. and Editorial Board of the journal Technology of Ancient Building and Landscape Garden.

I would like to express my thanks to Mr. Luo Zhewen, a specialist in cultural relics and ancient buildings for his handwriting of the title of this collection and Mr. Du Xianzhou for his "Preface" to this book.

He Junshou
January 1999

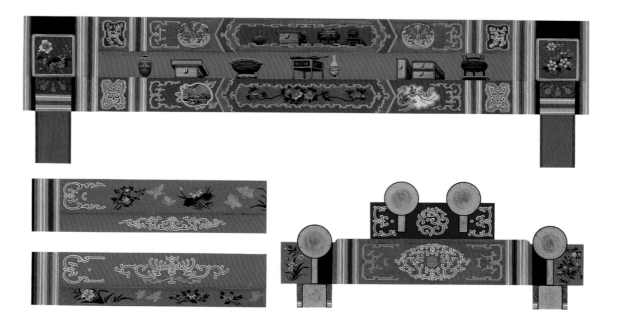

修订版附言

本图集是以改革开放后在《古建园林技术》期刊中陆续发表的中国建筑彩画样稿（该期刊特有的一个栏目）和一些专业单位及彩画专家几代人几十年的作品为底本，汇集一些科研成果而编印的。当时只限于内部交流，而未刊行于世。1993年为纪念《古建园林技术》创刊十周年，作者们再次整理编辑成册作为一份礼物献给杂志。随后很多专业人员、读者及有关各界知晓后，向编辑部和北京市古代建筑设计研究所建议编辑出版一本较为完整的专业图集，以适应更多人员的需要。1999年正值国庆五十周年之际，天津大学出版社积极倡议和精心策划，在古代建筑设计研究所和编辑部协作、赞助下，共同进行采访、搜集、组稿、评审并撰文，于当年将《中国建筑彩画图集》奉献给了读者。

图集出版五年来，一直得到读者和文物、建筑、美工专业人员的关注。其间不断来信或致电编辑人员和作者，一方面给予鼓励，肯定图集的价值和作用，同时诚恳热情地指出在用字、文词、版面方面存在的不足和疏漏。值此出版"修订版"之际，我们几位编辑又重聚一起认真采纳来自各方面的批评建议，对第一版作了详慎的校核，并斟字酌句予以更正。为此向专家学者和师哲们致以由衷的感谢和敬意！

本修订版进行了如下充实、调整。

1. 1999年为了迎接第二十届世界建筑师大会在中国召开，本书作为向大会献礼，编辑、制版、印刷等各阶段工作过于紧张，当时未能来得及由撰稿人和作者点校"清样"，致使印制的书籍中出现了几处不应有的错误。这次作了改正、调整。

2. 本版增加了一些新采集、挖掘的样稿及修复作品，弥补了原图集中官式建筑彩画主要品类的不足。如：浑金旋子彩画、宝珠吉祥草彩画等，共计50多项。在栏目、画种分类、序列排印上，全面作了必要的调整。

3. 此次我们经过研究，对中国建筑彩画的所用名称、用语、用字存在的不尽统一等问题进行了调整。同时还对本图集涉及引用的参考文献作了考证，取其合宜者作了修正。例如：①"枋心"统一更正为"方心"；②"混金"改用"浑金"；③"卡箍头"均用"掐箍头"；④将"贯套箍头"或"环套箍头"定为"观（guàn）头箍头"。

通过此次修订，我们力图做到比较完美，虽经努力，但由于参与人员水平所限，仍会有不尽如人意之处，尚请各行方家给予指教。

何俊寿

2005年6月

目 录

明、清官式彩画的概况及工艺特征　　王仲杰 ………………1-6

明代彩画 ……………………………………………………… 15
- 北京智化寺明代金线旋子彩画　北京文物整理委员会 ……………… 15
- 北京智化寺明代金线旋子彩画　北京文物整理委员会 ……………… 16
- 北京智化寺万佛阁下层梵文天花　北京文物整理委员会 …………… 17
- 北京东四牌楼清真寺礼拜殿内檐包袱彩画　北京文物整理委员会 ……… 18
- 北京东四牌楼清真寺礼拜殿内檐柱彩画　北京文物整理委员会 ………… 19
- 北京磨石口法海寺山门外檐明代雅五墨彩画　杜恒昌 ………………… 20
- 北京磨石口法海寺山门脊部明代雅五墨彩画　高成良 ………………… 21
- 明代墨线点金旋子彩画　杜恒昌 …………………………………… 22

和玺彩画简述　　蒋广全 ……………………………………… 23
- 龙和玺彩画　边精一　高成良 ……………………………………… 24
- 龙和玺彩画　杜恒昌　高成良 ……………………………………… 25
- 龙和玺彩画　北京文物整理委员会 ………………………………… 26
- 金琢墨龙和玺彩画　杜恒昌 ………………………………………… 27
- 龙凤和玺彩画　边精一　杜恒昌 …………………………………… 28
- 凤和玺彩画　蒋广全 ………………………………………………… 29
- 龙草和玺彩画　蒋广全 ……………………………………………… 30
- 内檐梁架龙草和玺彩画　高成良 …………………………………… 31
- 龙草和玺彩画　张秀芬 ……………………………………………… 32
- 梵文龙和玺彩画　冯世怀 …………………………………………… 33
- 龙草反搭包袱彩画　冯世怀 ………………………………………… 34

旋子彩画简述　　赵双成 ……………………………………… 35
- 夔龙方心金琢墨石碾玉旋子彩画　蒋广全 ………………………… 36
- 龙锦方心金琢墨石碾玉旋子彩画　郑书本 ………………………… 37
- 龙锦方心金琢墨石碾玉旋子彩画　边精一 ………………………… 38
- 龙锦方心金线大点金旋子彩画　边精一　高成良 ………………… 39

金线大点金绘画方心旋子彩画　杜恒昌	40
龙锦方心墨线大点金旋子彩画　王立新	41
龙草方心墨线大点金旋子彩画　冯世怀	42
片金西蕃莲宋锦方心墨线大点金旋子彩画　高成良	43
片金西蕃莲宋锦方心墨线大点金旋子彩画　冯世怀	44
夔龙西蕃莲方心墨线小点金旋子彩画　北京文物整理委员会	45
夔龙西蕃莲方心雅五墨旋子彩画　冯世怀	46
夔龙西蕃莲方心雅五墨旋子彩画　冯世怀	47
一字方心雅五墨旋子彩画　杜恒昌	48
夔龙花卉方心雄黄玉旋子彩画　高成良	49
一字方心雄黄玉旋子彩画　赵双成	50
旋子彩画锦纹两则　王立新	51
旋子彩画锦纹两则　王立新	52

苏式彩画简述　杜恒昌　53

清中期方心式苏画(一)　园林古建公司	54
清中期方心式苏画(二)　杜恒昌	55
清中期方心式苏画(三)　杜恒昌	56
清中期方心式苏画(四)　杜恒昌	57
清中期方心式云秋木苏画(五)　蒋广全	58
清中期方心式苏画(六)　蒋广全	59
清中期方心式苏画(七)　园林古建公司	60
清中期包袱式苏画(八)　园林古建公司	61
清中期包袱式苏画(九)　蒋广全	62
清中期包袱式苏画(十)　蒋广全	63
清中期墨线海墁锦纹双蝠葫芦团花苏式彩画　蒋广全	64
清中期金线海墁流云黑叶花卉苏式彩画　蒋广全	65
清中期海墁式苏画　冯世怀	66
清晚期方心式苏画　杜恒昌	67
清晚期方心式苏画　冯世怀	68
清晚期方心式苏画　冯世怀	69
金琢墨方心式苏画　边精一　杜恒昌	70
清晚期包袱式苏画　高成良	71
包袱式苏画　杜恒昌	72
汉瓦箍头包袱式苏画　边精一	73
清晚期包袱式苏画　蒋广全　赵双成	74
清晚期包袱式苏画　冯世怀	75
清晚期挂檐方心式苏画　高成良	76
掐箍头搭包袱苏画　杜恒昌	77
掐箍头彩画　赵金城	78

掐箍头彩画　赵金城	79
金线垂花门方心苏画　蒋广全	80
墨线垂花门方心苏画　蒋广全	81
包袱聚锦彩画　杜恒昌	82
落墨搭色人物包袱彩画　蒋广全	83
包袱聚锦彩画　高成良	84
包袱彩画　蒋广全	85
包袱聚锦彩画　杜恒昌	86
聚锦彩画　高成良	87
聚锦壳　高成良	88
聚锦壳　高成良	89
游廊包袱式苏画图卷（1）北京古代建筑公司集体创作	90
游廊包袱式苏画图卷（2）北京古代建筑公司集体创作	91
游廊包袱式苏画图卷（3）北京古代建筑公司集体创作	92
游廊包袱式苏画图卷（4）北京古代建筑公司集体创作	93

宝珠吉祥草彩画、海墁（斑竹座）彩画简述　冯世怀……94

| 清代宝珠吉祥草彩画　蒋广全 | 95 |
| 清代斑竹纹海墁彩画　蒋广全 | 96 |

清代官式彩画檩枋大木以外从属构件简述　蒋广全……97

椽头、飞头、椽子、望板彩画　蒋广全	98
斗拱彩画　边精一	99
斗拱彩画　边精一	100
垫拱板彩画　杜恒昌	101
片金坐龙天花　杜恒昌	102
片金坐龙天花　蒋广全	103
片金升降龙天花　蒋广全	104
片金双凤天花　蒋广全	105
团鹤天花　蒋广全	106
双鹤天花　蒋广全	107
夔龙岔角六字真言天花　蒋广全	108
六字真言天花　冯世怀	109
坐夔龙天花　蒋广全	110
攒退硬夔龙天花　蒋广全	111
福寿天花　蒋广全	112
百花天花　蒋广全	113
彩龙天花　高业京供稿	114
片金西蕃莲天花　蒋广全	115
雀替彩画　蒋广全	116

地方彩画简述　王仲杰 …… 117
清晚期苏州地方彩画　高成良 …… 118
山西地方彩画　岳俊德等 …… 119
山西地方彩画　岳俊德等 …… 120
辽宁地方彩画　张世满　李海申 …… 121
黑龙江地方彩画　郑连昶 …… 122
黑龙江地方彩画　郑连昶 …… 123

新式彩画简述　高成良 …… 124
新式彩画——藻井　郭汉图 …… 125
新式彩画——墙边　郭汉图 …… 126
新式彩画——墙边　郭汉图 …… 127
新式彩画——墙边　郭汉图 …… 128
新式彩画——双龙、双凤天花　郭汉图 …… 129
新式彩画——灯花　赵双成 …… 130
新式彩画——灯花　杜恒昌 …… 131
新式彩画——灯花　赵双成 …… 132
新式彩画——灯花　蒋广全 …… 133
新式彩画——灯花　蒋广全 …… 134
新式彩画——灯花　罗翰秋 …… 135
新式彩画——灯花　杜恒昌 …… 136
新式彩画——柱　郭汉图 …… 137
新式彩画——柱　郭汉图 …… 138
新式彩画——柱　赵双成 …… 139
新式大木彩画　郑书本 …… 140
新式大木彩画　郭汉图 …… 141
新式大木彩画　杜恒昌 …… 142
壁画——八仙醉酒　冯世怀 …… 143
壁画——瑶池盛会　冯世怀 …… 144
壁画——太白醉酒　冯世怀 …… 145

实物照片 …… 146
（上）历代帝王庙景德崇圣殿明间内檐脊部彩画　蒋广全摄影 …… 146
（下）故宫景仁宫龙凤方心西蕃莲找头和玺彩画　蒋广全摄影 …… 146
（上）牌楼和玺彩画　马炳坚等摄影 …… 147
（下）牌楼和玺彩画　马炳坚等摄影 …… 147
（上）浑金做法旋子彩画　蒋广全供稿
（引自于倬云先生主编的《紫禁城宫殿》） …… 148
（下）北京八大处六处金龙方心墨线大点金旋子彩画　蒋广全摄影 …… 148
（上）北京法兴寺内檐墨线大点金旋子彩画　蒋广全摄影 …… 149
（下）北京法兴寺内檐墨线大点金旋子彩画（局部）　蒋广全摄影 …… 149

（上）历代帝王庙墨线大点金旋子彩画　蒋广全摄影 ……………………… 150
（下）文丞相祠墨线小点金一字方心旋子彩画　高成良摄影 …………… 150
（上）承德普宁寺西配殿内檐烟琢墨石碾玉旋子彩画　蒋广全摄影 …… 151
（下）承德普宁寺钟楼彩画修缮后　蒋广全摄影 ………………………… 151
　　　牌楼旋子彩画　马炳坚等摄影 …………………………………… 152
（右）一字方心雄黄玉彩画　蒋广全摄影 ………………………………… 152
（左）椽子、望板彩画　蒋广全摄影 ……………………………………… 153
（上）浑金做法龙雀替　蒋广全摄影 ……………………………………… 154
（下）墙边做法两例　蒋广全摄影 ………………………………………… 154
（上）椽、柁头刷饰　马炳坚等摄影 ……………………………………… 155
（下）椽、柁头彩画　马炳坚等摄影 ……………………………………… 155
（上）椽、柁头彩画　马炳坚等摄影 ……………………………………… 156
（下）掐箍头彩画　马炳坚等摄影 ………………………………………… 156
（上）垂花门包袱式苏画　马炳坚等摄影 ………………………………… 157
（下）包袱式苏画　马炳坚等摄影 ………………………………………… 157
（上）内檐苏画　马炳坚等摄影 …………………………………………… 158
（下）包袱式苏画　林其浩摄影 …………………………………………… 158
（上）上方心式苏画和下海墁式苏画　马炳坚等摄影 …………………… 159
（下）方心式苏画　马炳坚等摄影 ………………………………………… 159
（上）垂头彩画　马炳坚等摄影 …………………………………………… 160
（下）花板彩画　马炳坚等摄影 …………………………………………… 160
（上）清中期方心式苏画（局部）　马炳坚等摄影 ……………………… 161
（下）清中期方心式苏画（局部）　马炳坚等摄影 ……………………… 161
（上）清中期方心式苏画（局部）　马炳坚等摄影 ……………………… 162
（下）清晚期包袱式苏画（局部）　马炳坚等摄影 ……………………… 162
（上）清晚期苏画（局部）　马炳坚等摄影 ……………………………… 163
（下）清晚期苏画（局部）　马炳坚等摄影 ……………………………… 163
（上）清晚期抱头梁、穿插枋苏画　马炳坚等摄影 ……………………… 164
（下）清晚期抱头梁苏画　马炳坚等摄影 ………………………………… 164
（上）清晚期游廊内檐苏画　马炳坚等摄影 ……………………………… 165
（下）清晚期游廊内檐苏画（局部）　马炳坚等摄影 …………………… 165
（上）清中期方心式苏画（局部）　马炳坚等摄影 ……………………… 166
（下）八角亭内檐苏画（局部）　马炳坚等摄影 ………………………… 166
（上）苏式彩画包袱（一）线法山水　马炳坚等摄影 …………………… 167
（下）苏式彩画包袱（二）线法人物　马炳坚等摄影 …………………… 167
（上）苏式彩画包袱（三）线法山水　马炳坚等摄影 …………………… 168
（下）苏式彩画包袱（四）线法人物　马炳坚等摄影 …………………… 168
（上）苏式彩画包袱（五）线法山水　马炳坚等摄影 …………………… 169
（下）苏式彩画包袱（六）线法人物　马炳坚等摄影 …………………… 169
（上）苏式彩画包袱（七）线法人物　马炳坚等摄影 …………………… 170
（下）苏式彩画包袱（八）富贵白头　马炳坚等摄影 …………………… 170
（上）苏式彩画包袱（九）花鸟　马炳坚等摄影 ………………………… 171

（下）苏式彩画包袱（十）花鸟　马炳坚等摄影 ... 171
（上）苏式彩画包袱（十一）松鹤延年　马炳坚等摄影 ... 172
（下）苏式彩画包袱（十二）花鸟　马炳坚等摄影 ... 172
（上）苏式彩画包袱（十三）百鸟朝凤　马炳坚等摄影 ... 173
（下）苏式彩画包袱（十四）海宴河清　马炳坚等摄影 ... 173
（上）苏式彩画包袱（十五）富贵满堂　马炳坚等摄影 ... 174
（下）苏式彩画包袱（十六）玉堂富贵　马炳坚等摄影 ... 174
（上）苏式彩画包袱（十七）三国人物故事　马炳坚等摄影 ... 175
（下）苏式彩画包袱（十八）三国人物故事　马炳坚等摄影 ... 175
（上）苏式彩画包袱（十九）三国人物故事　马炳坚等摄影 ... 176
（下）苏式彩画包袱（二十）三国人物故事　马炳坚等摄影 ... 176
（上）苏式彩画包袱（二十一）聊斋人物故事　马炳坚等摄影 ... 177
（中）苏式彩画包袱（二十二）三国人物故事　马炳坚等摄影 ... 177
（下）苏式彩画包袱（二十三）落墨山水　马炳坚等摄影 ... 177
（上）苏式彩画包袱（二十四）包公案人物故事　马炳坚等摄影 ... 178
（下）苏式彩画包袱（二十五）山水人物　马炳坚等摄影 ... 178

墨线图 ... 179

龙草和玺（一）　王仲杰　蒋广全　冯世怀　张秀芬 ... 179
龙草和玺（二）　王仲杰　蒋广全　冯世怀　张秀芬 ... 180
龙草和玺（三）　王仲杰　蒋广全　冯世怀　张秀芬 ... 181
龙草和玺（四）　王仲杰　蒋广全　冯世怀　张秀芬 ... 182
（左上）早期行龙　赵双成 ... 183
（右上）早期升龙　赵双成 ... 183
（左下）早期降龙　赵双成 ... 183
（右下）早期坐龙　赵双成 ... 183
（上）晚期行龙　赵双成 ... 184
（左中）晚期升龙　赵双成 ... 184
（左下）晚期降龙　赵双成 ... 184
（右）晚期坐龙　赵双成 ... 184
（上）夔行龙　赵双成 ... 185
（左中）夔升龙　赵双成 ... 185
（左下）夔降龙　赵双成 ... 185
（右）夔坐龙　赵双成 ... 185
（左上）行凤　赵双成 ... 186
（右上）升凤　赵双成 ... 186
（左下）降凤　赵双成 ... 186
（右下）团凤　赵双成 ... 186
（上）观头箍头　赵双成 ... 187
（下）各式箍头　蒋广全 ... 187
龙锦方心金线大点金旋子　蒋广全 ... 188
龙锦方心金线大点金旋子　蒋广全 ... 189

条目	作者	页码
旋子彩画梁架示意图	蒋广全	190
金线大点金旋子加包袱	蒋广全	191
小式旋子彩画	王仲杰	192
旋子彩画找头	蒋广全	193
旋子彩画找头	蒋广全	194
（上）平板枋纹饰	赵双成	195
（下）池子纹饰	蒋广全	195
宝瓶、纹饰	蒋广全	196
异兽	赵双成	197
包袱式苏画、方心式苏画、海墁式苏画	蒋广全	198
苏式彩画包袱边	蒋广全	199
苏式彩画包袱心	蒋广全	200
包袱心两则	冯世怀	201
包袱心两则	冯世怀	202
包袱心两则	冯世怀	203
包袱心两则	冯世怀	204
卡子	蒋广全	205
聚锦壳、流云	蒋广全	206
（上）柁头、	蒋广全	207
（下）柁头帮	赵双成	207
枋底切活	赵双成	208
坐龙天花、金莲水草天花	蒋广全	209
双凤天花、双鹤天花	蒋广全	210
西蕃莲天花、五福（蝠）捧寿天花	蒋广全	211
（上）宝珠吉祥草彩画	王仲杰	212
（下）斑竹纹海墁彩画	蒋广全	212
蟠龙柱、新式柱纹三则	赵双成	213
新式彩画灯花四则	赵双成	214

王仲杰

明、清官式彩画的概况及工艺特征

彩画是我国古建筑的重要组成部分，它伴随着古建筑的发展而发展。彩画的演变也和古建筑的其他部分一样，都有一个由简单到复杂、由低级到高级的进化过程。由于年代久远，早期的实物例证难以保存下来，因此彩画的早期状况已经难以全部弄清，后人只能了解个梗概。唐、宋、辽、金各代留的遗存很少，可谓"凤毛麟角"，但总算还有可见的实物，又加之有北宋官修的《营造法式》相对应，这就为后人认识这一历史阶段的彩画状况提供了例证和文献依据。元、明、清三代（特别是清代）遗存的历史原迹十分丰富，并有清雍正朝官颁的《工程做法》传世。更为可喜的是，如北京等一些文化古城从事古建筑彩画修缮的匠师间，还沿袭着一套比较完整的传统做法。三者并存，为后人整理研究这一历史阶段官式彩画的文化和技术内涵奠定了物质基础。

为了配合本书的出版，我冒昧地就明、清时期官式彩画的概况作些肤浅的分析，实属引玉之砖。

明、清两代，尤其是清代是我国建筑彩画发展史上最活跃、硕果最丰盛的时期。新的品种不断涌现；题材不断扩大；表现手段不断丰富；法式规矩更加严密规范；等级层次更加严明、清晰。这些成就除了有前朝前代所奠定的基础外，客观条件也为彩画的发展提供了文化和物质方面的基础，比如说颜料的进一步丰富，特别是金箔产量的剧增，这就为彩画朝着富丽化发展提供了条件。我国兄弟民族间文化的进一步交融，文人画、民俗画的普及，以及外来文化的传入都为彩画的快速发展提供了题材和工艺方面的新来源。

这一时期的彩画，从大的方面归纳，有官式做法和地方做法两种。前者是当时建筑管理部门按照当时的等级制度和工料限额直接组织官式工匠制作的一种定型的彩画。它的服务对象是皇家御用建筑、王公大臣府第、敕建庙宇及京城衙署等。后者是指民间工匠在不违背当时等级制度的前提下，施绘于地方衙署、庙宇和民居建筑上的一类比较活泼自然、不拘泥程式的彩画。两者做法虽然不尽一致，但又是互通互补的。地方做法总是尽力效法官式做法，官式做法也不断地吸取地方做法中可用成分来充实自己。从总的方面来讲，代表这个时期最高水平、最具权威性的当属官式做法。

一、明代官式彩画

由于遗存的旧物较少，我们已无法认识这一阶段彩画的全貌。明代官式彩画究竟有几类至今还是个谜。就我们目前所见到的实物例证来看仅有两类：其一是"旋子彩画"（究竟定为何名更为准确，留待以后研究，为了叙述方便暂且套用"旋子"一名）。其二是绘于皇家园囿建筑之上的、近似于清代中期官式苏画的龙纹方心、锦纹找头彩画。现存实物绝大部分为旋子彩画，第二类彩画只存一例。故本文只着重于旋子类彩画作些分析。

明代官式旋子彩画是从元代同类彩画演变形成的，是有物可证明的。梁枋大木部分彩画是在继承的基础上，改变了元代梁枋大木彩画构图自然的风范，形成了构图严谨的风格。变化的突出之处是方心和找头两个部位。方心的端头造型固定为⁊型。方心内一般不施绘细部纹饰，只平涂颜色，即所谓素方心。这一做法在前代是不多见的。找头的纹饰已基本固定为旋花造型，"一整两破"成为基本形式。旋花的形象虽然已经基本图案化，但局部仍然保留着写生作画的痕迹。例如有的旋花心绘莲座，其上再绘莲实、莲蕊。这时的找头虽然已经以旋花作为主体纹饰，但依然穿插使用如意头等纹饰，以避免一种纹饰反复出现的弊病。箍头在明代官式彩画中已经定型，无论构件长短均以箍头作为端头的收尾。

明代施用旋子彩画的范围，比起元代和清代都要大些，最常见的是在木构件表面用颜料涂绘的旋子彩画。除此之外，还有用烧制琉璃材料塑出纹饰而成的琉璃旋子彩画（如明十三陵、湖北武当山等处的琉璃门等），在铜质构件表面用线刻的旋子彩画（如武当山金顶），以及在石质构件表面雕刻的旋子彩画等等。

以上所举的几种不同质地的旋子彩画，尽管底材不同，所在的地理位置跨度很大，但它们的纹饰几乎是一模一样，如出一范。由此可以说，明代官式旋子彩画已经完全有了定式，已经有了一整套完整的法式规矩用以控制操作了，在此之前历代尚未达到如此高的水平。明代官式彩画中斗拱的装饰，已经演变成青绿色退晕做法，不再其上施绘细部花纹。明代官式彩画的天花构图已经确定了方、圆鼓子造型，圆鼓内的细部花纹比较自然，有西蕃莲纹、佛梵字等等。

（一）明代官式彩画的设色和工艺特点

明代官式彩画的设色和工艺特点，如同它的纹饰一样具有鲜明特色。它没有大片为描绘写生画的白色画面，红色使用也很少。基础颜色是青色和绿色两大主色。明代官式彩画之所以给人们淡雅、简洁之感，全赖于选择颜料之妙和工艺的高超。

1.颜料。此时彩画所使用的颜料全部为国产的矿物质材料——石青、石绿、银朱等，其色感近似国画颜料，艳而不浮。用这些颜料绘制的彩画犹如一幅立体的青绿重彩装饰画。近几十年古建界也曾在多处明代古建筑上补绘了明代官式彩画，色调方面总是不尽如人意。究其原因，所使用的颜料为当代的巴黎绿、群青等颜料。这些颜料较石青、石绿鲜艳有余，但少了些深沉之感。由此可见明代官式彩画的淡雅之感是与其所使用的颜料成分分不开的。

2.色调。在暖调颜色和冷调颜色的组合方面也是颇有特色的，如梁枋彩画在一片青绿基色之中用红色装点花心的莲座，顿时把莲花这一主题突出出来，同时也打破了画面的呆板。用金也是如此，它不是遍施金箔，把彩画装饰得珠光宝气，而是在最突出的部位，如花心、花蕊或菱角地点饰一下，往往又不十分对称，以避雷同之弊。

3.做法。明代官式旋子彩画主要工艺特征是：无论彩画的等级高低，一律采用退晕做法，晕色的退法均由浅色入手，逐层加深，且色阶的宽度差别不大。就目前的实物看，还没有出现以白色为起点勾勒白线的做法，其最浅的色阶乃是浅青色或浅绿色。此时的彩画之所以给人以色调柔和之感，是因为它没有用反差很大的白色。除了梁枋彩画之外，其余构件的装饰方法也是和梁枋彩画做法是一致的，于是构成了淡雅、柔和的时代风格。

（二）明代官式旋子彩画的等级制度

关于明代官式旋子彩画的等级制度，目前不是很清楚。主要原因是实物例证不足，文献记载不详。现存的实物中最高等级是后妃居住的殿宇，其次是皇宫内偏僻院落中的殿堂及一些祭祀、宗教建筑之上的彩画，缺少像皇帝御用建筑的彩画实例，也缺少府第衙署和民居的实例，这就增加了分析的难度。现在我们只能根据这些存在的实际资料作推断性的分析。从现存的例证看，旋子彩画是此时的主要品种。我们虽然不可能见到那时皇帝登基理政的金銮宝殿——奉天殿彩画的样子，但是可用同期供奉帝君神佛的建筑彩画作旁证。府第衙署虽无实证，但在庙宇的彩画中可以找到一些影子。由此推断明代官式旋子彩画的等级是以绘制的精细程度和用金箔多少而划分的。其大致可划分为三个等级。

1.金线大点金彩画。这一等级的彩画其方心多为无花的素方心，最多在其内绘些龙纹。主体框架线全为沥粉贴金做法，花心、菱角地也为沥粉贴金做法。其天花做法与梁枋相一致的均为金线，细部纹饰局部点缀金色。斗拱为青绿退晕做法。这是当时的最高等级彩画，适用于皇宫内的主要殿宇和重要坛庙的主殿。

2.墨线点金彩画。这一等级的彩画方心不施绘细部纹饰，主体框架线一律为墨线，找头中的花心、菱角地、如意头等部分点缀金色（因为点缀金色的部位和数量非常自然，我们不便称其为什么点金，只好笼统称其为点金彩画）。其天花的用金量相应减少，斗拱做法同大点金彩画，它是当时的中等级做法，适用范围也比较宽，如皇宫内后、妃居住的殿堂和比较重要的殿宇及各种宗教寺庙的主要殿宇。

3.不点金彩画。这一等级的彩画适用于一般建筑。

关于苏式彩画，有专家推论是明永乐营建北京宫殿时，随着苏州工匠进京带进京师，但因缺少实物例证，今天已无法认识其具体形象。

明代的地方做法彩画现存也是很少的，从仅存的实例看，大致为两部分。一种是北方寺庙殿宇上的旋子彩画。它的纹饰结构、设色方法基本上同于官式做法，只是不那么规范，比较自然，从其现状分析这些彩画是官式彩画的摹拟品。

另一种是江南民居房屋上的彩画。它的纹饰构图多以包袱为主体纹饰，细部纹饰、设色都比较纤细、淡雅，具有江南特有的风韵。

二、清代官式彩画

彩画也如其他艺术作品一样，不可能随着封建王朝的更迭而随之骤变。它的变化也是有一个渐变的过程。清代适用于彩画的颜料不断丰富，美术领域的空前活跃和统治者追求生活环境更加富丽堂皇，促使了官式彩画的飞速发展。

清代官式彩画的类别比起明代丰富得多。从纹饰的主体框架构图和题材方面分类，至少可以划分为五大类，即：和玺类、旋子类、苏式类、吉祥草类和海墁类。

（一）和玺彩画

和玺彩画是清代的最高等级彩画，它的形成年代比起其他类别的彩画要年轻得多。明代中期以前尚无此种彩画，它的出现和成型的时间大约在明末清初之际。和玺彩画是在旋子彩画的基础框架上演化出来的，它保持了旋子彩画的基本格局：箍头、找头—方心—找头、箍头的三大段落，只是把找头两端的60度角皮条线、岔口线删去，换上三个横向排列的莲瓣形边框，再去掉旋花。和玺彩画纹饰方面的另一个突出特点是方心、找头、盒子及平板枋、垫板等构件不施绘锦纹和花卉，而遍绘龙纹、凤纹、西蕃莲纹及吉祥草纹。在设色方面，和玺彩画不同于其他类别彩画，主体框架线一律为沥粉贴金做法，不采用墨线做法。细部纹饰大部分也为沥粉贴金做法。

从细部纹饰题材方面分析，和玺彩画主要有以下五种绘法。

1. 龙和玺。即梁枋大木中的方心、找头、盒子及平板枋、垫板、柱头等构件全部绘龙纹。彩画界称这种彩画为"金龙和玺"或"五龙和玺"。所谓五龙和玺是指檩子为一龙，垫拱板为一龙，平板枋为一龙，大额方为一龙，柱头为一龙。这种记数方法不甚缜密，施绘龙纹的构件不限于以上所举的这些构件。总之这种和玺是以龙纹为基础纹饰，还是称其为"龙和玺"更为贴切。

2. 凤和玺。即梁枋大木中的方心、找头、盒子及平板枋、垫板、柱头等构件全部绘以凤纹。

3. 龙凤和玺。是以龙纹、凤纹相匹配组合的一种和玺。

4. 龙凤方心西蕃莲灵芝找头和玺。即方心和盒子绘以龙纹、凤纹，找头内分别绘以西蕃莲纹和灵芝纹。

5. 龙草和玺。即梁枋大木的方心、找头、盒子及平板枋、垫板等构件采用龙纹与吉祥草纹互换排列的方式组合的一种和玺。

和玺彩画除了上面所举的几种纹饰组合的和玺外，在佛教建筑上所绘的和玺彩画中常绘以表示宗教教义的梵文、宝塔和莲座等纹饰。

和玺彩画的装饰中心部件是上面所举的梁枋大木构件，与之相关联的其他构件也是相当精致的。如高等级的和玺彩画，其椽子望板可以遍绘纹饰，角梁底面可以绘龙纹。无论大木绘哪种和玺彩画，其斗栱的边线一律采用金箔装饰。

从设色和工艺方面看，和玺彩画又是一种比较简单的彩画，几乎所有纹饰都是采用沥粉贴金方法制作的。只是盒子岔角和方心、找头中的云纹采用"攒退"做法，底色都为单色平涂。早期和玺只在主体框架线的一侧绘一条浅青色或浅绿色线条。晚期和玺在主体框架线的一侧加绘一条晕色，并将浅青色、浅绿色的线条改为白色，借以显示其韵味。和玺彩画的装点妙处主要在于金色和青绿底色的反差作用上。在一片凝重的底色上面，遍绘耀眼的金色纹饰，把建筑点染得金碧辉煌。一般和玺只贴一种金箔，高等级的和玺多采用贴深浅两色金箔（即库金箔和赤金箔），进一步增加深邃感。

和玺虽属高类别的彩画，但这类彩画的等级层次还是十分严明的。龙和玺是和玺类的第一等，只适用于皇帝登极、理政的殿宇和重要坛庙的主殿。龙凤和玺、龙凤方心西蕃莲灵芝找头和玺是和玺类的第二等，仅次于龙和玺，适用于帝后寝宫和祭天建筑的主殿（如天坛祈年殿）。凤和玺适用范围较窄，只用于皇后寝宫和祭祀后土神的殿宇（如地坛的皇祇室）。从等级方面分析，此种和玺也应属于二等。龙草和玺是和玺类中最低的一等，适用于皇宫的重要宫门和中轴线上配殿、配楼及重要寺庙的主要殿堂。

（二）旋子彩画

清代的旋子彩画是在明代旋子彩画的基础之上演变而成的。这类彩画品种繁多，使用广泛，是清代官式彩画中的一个主要类别。从其纹饰组合、设色和做法三个方面分析大致有八种，即：

(1) 浑金旋子彩画；

(2) 金琢墨石碾玉旋子彩画；

(3) 烟琢墨石碾玉旋子彩画；

(4) 金线大点金旋子彩画；

(5) 墨线大点金旋子彩画；

(6) 小点金旋子彩画；

(7) 雅五墨旋子彩画；

（8）雄黄玉旋子彩画。

旋子彩画的基础构图是一致的，檩枋大木的两端绘箍头，大开间的檩枋在箍头的内侧括出一个近方形的盒子，盒子的内侧再增设一条箍头。在檩枋的中部括出一个占构件全长1/3的长方形画框，称为方心。在方心和箍头之间的这部分称之为找头。其间用三层或两层花瓣所组成的大团花（旋花）若干朵置于其中，最常见的是由一个整团花和两个半朵团花相组合的形式（这种组合形式称之为"一整两破"）。找头的长短随着构件的长短而相应调整，其间的旋花也相应变化。大开间的找头可以安排若干个"一整两破"，小开间的找头可以将旋花局部重叠起来或取其一部分。因团花的最外一层花瓣呈◎形，有旋转的意境，故取名旋花。旋花是这类彩画的基础纹饰，因此这种彩画也就称之为旋子彩画。

旋子彩画不论其等级高低，找头中的旋花纹饰是不能改变的。而方心部分和盒子部分的细部花纹可随着等级高低而变化。方心内纹饰从高到低的层次是：龙纹、龙凤纹、凤纹、锦纹、夔龙纹、卷草纹、花卉等。有的旋子彩画方心内只用黑色压底，甚至可不绘任何纹饰而裸露底色。所谓"盒子"即两条箍头之间其长度与额方的高度相近的一个方形体。高等级的旋子彩画在其间绘出一个由八条弧线所组成的近圆形画框，其内绘龙纹、西蕃莲纹或异兽等纹饰，这种绘法称之为活盒子。低等级的旋子彩画盒子部分不括出圆形画框，只在方形体内用四个花瓣形栀花纹组合成几何纹，这种绘法称之为栀花盒子或称之为死盒子。

现就纹饰组合和设色等方面的变化，简要地介绍旋子彩画不同品种的具体做法。

1.浑金旋子彩画。 方心内不布置细部纹饰，盒子也只绘栀花纹。所有纹饰皆用沥粉线显示，整个画面不敷色彩，全部贴金箔。

2.金琢墨石碾玉旋子彩画。 方心、盒子内绘龙纹、夔龙纹、凤纹等纹饰。主体框架线及细部纹饰均用沥粉做法，然后遍贴金箔。主体框架线及旋花全部为青绿叠晕做法。

3.烟琢墨石碾玉旋子彩画。 方心、盒子内的细部纹饰基本上与金琢墨石碾玉旋子彩画相同，也有采用素方心和死盒子的实例。主体框架线及找头中的旋眼、菱角地、栀花心和方心、盒子内的纹饰均采用沥粉贴金做法。旋花、栀花的边线皆为墨线，主体框架线及旋花全部为青绿叠晕做法。

4.金线大点金旋子彩画。 方心内的纹饰基本上以龙纹和锦纹为主，两种纹饰匹配组合，专业术语称之为"龙锦方心"，也有只用龙纹的。盒子内的纹饰多采用龙纹和西蕃莲纹匹配组合。主体框架线及找头中的旋眼、菱角地、栀花心和方心的龙纹，锦纹中的部分花纹及盒子内的龙纹、西蕃莲纹均采用沥粉贴金做法。主体框架线为青绿色叠晕做法，旋花和栀花只在青绿底色之上用黑色勾勒边线，然后沿边线内侧描一道白色粉线。

5.墨线大点金旋子彩画。 方心内多数不布置细部纹饰，个别实例也有绘龙纹、锦纹的。盒子多为"死盒子"，旋眼、菱角地、栀花心沥粉贴金（如方心内绘龙纹、锦纹或西蕃莲纹全部或局部贴金箔）。主体框架线及旋花、栀花皆为墨线，边线一侧描绘一道白色粉线。

6.小点金旋子彩画。 方心内多绘夔龙纹（攒退做法）和绿地墨叶子花卉，两种纹饰匹配组合，专业术语称之为"夔龙、花卉方心"。也有不布置细部花纹的素方心做法，盒子一般为"死盒子"，贴金只限于旋眼和栀花心两部分。主体框架线为墨线，旋花（包括菱角地在内）只在青绿底色之上用黑色勾勒成纹，边线一侧描绘一道白色粉线。

7.雅五墨旋子彩画。 纹饰与小点金相同，只是不贴金箔。细部做法也与小点金相同。

8.雄黄玉旋子彩画。 其纹饰与低等级的旋子彩画是一致的，多为素方心、死盒子做法。这种彩画在设色方面与其他旋子彩画差别极大。它的主体颜色不是青、绿色，而是一律以雄黄色作底色，然后用白色勾绘主体框架线和旋花、栀花纹。凡表示为青色部分只沿边线叠一道浅青色的晕色，其上再绘一道青色浅线。凡表示为绿色部分也和表示青色一样，以深浅绿色表示。雄黄玉彩画一般不贴金箔。

旋子彩画中大木梁枋以外的诸种构件装饰是与梁枋彩画相匹配的。金线大点金以上的彩画，其椽头多为龙眼宝珠，飞头均为片金万字纹饰。斗拱的边线为金色，角梁、霸王拳、三岔头、将出头多为金边金花拉晕色的做法。雀替、花板等为金大边，其木雕草为烟琢墨攒退做法。墨线大点金、小点金一类中等级旋子彩画在这些构件装饰上不甚严格。

突出的部位是角梁和霸王拳，有采用金边黑老做法的，也有采用黑边黑老做法的。雀替、花板的雕草多为渲染白粉做法，也有采用攒退做法的，但这一等级的斗拱皆为黑边做法。

雅五墨是旋子彩画中的最低等级，飞头为黑万字，椽头为虎眼宝珠做法，斗拱、角梁全为黑边黑花，雀替、花板等的大边为黄色，木雕草全为渲染白粉做法。

旋子彩画是清代官式彩画中的第二类。在实际应用上大致可分为四个范围。

其一是皇宫、皇家园囿中的次要建筑，如一般殿堂、门庑、值房等多采用这类彩画。其中使用最多的是金线大点金和墨线大点金两种。个别比较重要的殿堂亦采用金琢墨石碾玉或烟琢墨石碾玉。值房一类低等级建筑多采用小点金或雅五墨。

其二是皇宫内外祭祀祖先的殿堂（如奉先殿、太庙等），帝后陵寝的主体建筑采用这类彩画中的高等级品种，如故宫奉先殿内所绘的是浑金旋子彩画，太庙诸殿绘的是烟琢墨石碾玉旋子彩画。清东、西陵的主体建筑也绘的是烟琢墨石碾玉旋子彩画。

其三是重要祭祀坛庙的次要建筑及一般庙宇和王府等也都采用旋子彩画。

其四是雄黄玉旋子彩画是一种专用的彩画，主要用于庖制祭品的建筑装饰上，如帝后陵寝及坛庙的神厨、神库等。有时也有例外，如北海阅古楼本是贮存法帖的建筑，也采用此种彩画。

（三）苏式彩画

苏式彩画是装饰园林建筑的一种彩画，它源于江南水乡苏州一带，传至北方进入宫廷即成为官式彩画中的一个重要品种。目前，我们所见到早期官式苏画，大部分是乾隆时期的遗物。从其总体构图至细部纹饰分析已经是官式化、北方化了，很难再看出苏州彩画的痕迹。只是沿用其名，而实际完全演变成一种独具特色的彩画。清代官式苏画大体上可分为两个阶段，即早中期官式苏画和晚期官式苏画。

1. 早中期官式苏画。这个时期的苏画从其构图上分析，大约可分为三种，即方心式、包袱式、海墁式。

方心式苏画的主体框架线与旋子彩画的主体框架线是近似的。换句话来说，它借用了旋子彩画主体框架线，着重在找头部分作了些变动，删去了旋花，换上锦纹、团花、卡子、聚锦一类图案。方心部分基本未变，仍然绘龙纹、凤纹、西蕃莲等纹饰，最多在其间绘些博古或写生画。

包袱式苏画的主体构图与方心式的主体构图不同之处是删去了方心，在其位置改为一个半圆形的画框，覆在檩、垫板和额枋的中部，因其形象酷似一个下垂的圆形花巾，故称其为"包袱"。以此构图的彩画，也随之称为"包袱式苏画"。这种彩画找头部分的画作与方心式基本一样，包袱内多绘"寿山福海"一类的吉祥图案或锦纹。

海墁式苏画的梁方两端只保留箍头，其间的方心、包袱、池子、找头一律删去，不设任何画框，使其成为一个开阔的画面，最多在箍头以里绘上一对卡子，其上遍绘卷草纹、蝠磬纹或黑叶子花卉。

从细部纹饰题材方面分析，这时苏画的主题依然与和玺、旋子彩画一样，是以龙纹、凤纹为主，其次是吉祥图案、写生画所占的比例很小。这种构图决定了它的绘制工艺与和玺、旋子彩画无大的区别，遍刷青绿底色，沥粉贴金，只是攒退活的比重要大些，局部增加些写生画。

这时的官式苏画生活气息较少，也不甚活泼，完全是一种工整、严肃的殿式彩画。目前，尚未发现清代早中期富庶人家宅院的苏式彩画实例，这部分彩画空间究竟是什么样子，已经难搞清楚了。

2. 晚期苏式彩画。这时的苏式彩画与早中期苏式彩画相比较，在类别方面没有大的变化，依然是三种格式，即包袱式、方心式和海墁式，但在细部纹饰方面的变化是非常明显的。包袱的边框造型由花边边框和烟云边框并存时代基本上演变为单一的烟云边框时代。箍头从原来以素箍头为主的做法，进化为以万字纹、回纹为主的做法。而素箍头退至第二位。找头部分已不再布置锦纹。此时的包袱式苏画（包括方心式苏画在内）最主要的变化，在于包袱、方心、池子内的纹饰从早中期以图案画为主。演变成了以写生画为主。连同"聚锦"内的写生画，这些都构成了晚期苏式彩画的主体画面。

写生画包括有故事情节的人物画、绚丽多姿的花鸟画、赋有诗情画意的楼阁画及山水画和水墨画等等。写生画表示的含意虽不如吉祥图案那样直接，但也都包含着吉祥寓意。方心式苏画晚期特征与包袱式是一致的。海墁式的主体框架构图与早中期相比无大变化，只是出现了流云纹和爬蔓花卉，并以流云纹和黑叶子花卉为主要题材。

大木以外诸种构件的装饰是与其大木彩画的等级相匹配的。凡大木之上苏画是金线、片金、金琢墨做法的，其椽头、飞头为片金、金边做法。如施绘彩画的建筑系有斗拱的大式建筑，其斗拱的边线也随之为金边。凡大木之上苏画是黄线或黑线做法的，椽头、飞头等部位做法也随之递减为黄边或黑边。

苏式彩画的等级划分比较困难，它不像和玺彩画与旋子彩画那样，细部纹饰都有严明的定制，甚至排列顺序都不能颠倒，其用金部位和用金多少都有详尽的规矩。苏式彩画在纹饰和工艺方面都是比较自由的。尤其是清代晚期更是这样。就我们现在所见到的早中期苏式彩画大致可分为两个品级：①高等级苏式彩画的主体线路为金线，龙纹、凤纹、花团、锦纹、卡子等细部纹饰为片金或金琢墨攒退做法；②低等级的苏式彩画的主体线路为墨线，卡子、蝠磬、卷草等细部纹饰基本为烟琢墨攒退做法，有时也在局部点缀一些金色。

晚期苏式彩画大致可划分为三个品级：①高等级苏式彩画的主体线路为金线，箍头、卡子等细部纹饰为金琢墨攒退做法，方心、包袱、聚锦、池子内的写生画以楼阁山水、金地花鸟为主；②中等级的苏式彩画的主体线路为金线，箍头、卡子等细部纹饰多为片金或烟琢墨攒退做法，方心、包袱内的写生画相应降低；③低等级的苏式彩画的主体线路为黄线，箍头多为素箍头，卡子等纹饰均为烟琢墨攒退做法。方心、包袱内的写生画进一步降低。

苏式彩画的等级关系难于划分。一般来讲，主体建筑采用高等级的彩画，次要建筑采用低等级的彩画。但有一点是明确的，代表皇权的龙凤纹在皇家御用建筑以外是不能采用的。

（四）吉祥草彩画

清代早期还存在一种构图别致、色彩炽烈的官式彩画——宝珠吉祥草彩画。其构图特征是在梁枋两端设箍头，在梁枋的中部绘一个以三个宝珠为核心、周围衬卷草纹的大型花团，从构件的底面向两侧铺展，犹如一个硕大的"反搭包袱"。侧面箍头以里的上端各绘一个由卷草纹所组合的岔角形纹饰。檩桁构件窄而长，只在其中部绘一个以三个宝珠为核心、两侧衬以卷草纹的纹饰。靠近箍头处各绘一组卷草（此构图近似中后期的公母草造型）。其设色方法是以朱红色作为通体底色，卷草纹为金抱瓣青绿色攒退做法，三个宝珠的外框沥粉贴金，用青绿色装点宝珠。整个画面简洁粗犷，色彩火炽。

这类彩画在清代早期是一个独立品种，多用于皇宫的城门和皇帝陵寝建筑，中期以后渐渐地与其他彩画相互融合，以新的形式出现，不再单独存在了。

（五）海墁彩画

这里所说的海墁彩画不是上面所介绍的苏式彩画中的海墁式苏画，而是整座建筑（包括连檐、椽望、上架大木和下架大木装修在内）遍绘一种纹饰的彩画。它极可能产生于清代晚期。从这类彩画的纹饰和设色方面分析大约有以下三种做法。

其一是遍绘斑竹纹（匠师称之为斑竹座），即在大小构件的表面绘出与构件平行的一排排竹杆形纹饰，犹如每个构件是用若干根竹杆拼攒而成的。在梁方等大型构件的两端用细竹杆纹组合成箍头，在构件的中部用细竹杆纹组合成盘长纹、福寿字等纹饰。竹杆绘成嫩竹色和老竹色，在竹节处点染斑纹，意在模仿天然斑竹。这种彩画意境甚雅，追求一种自然之美。

其二是所有构件均以绿色或淡黄色作为底色，其上绘串枝牡丹或藤萝等一类爬蔓花卉，将建筑装扮成举目皆花的环境。

其三是上架大木均以青色作为底色，其上遍绘彩色流云纹。

海墁类彩画立意新颖，但未被广泛使用，除了斑竹纹偶有用之，其余两种已不多见。

清代官式彩画是官式建筑的主要等级标志之一，在流传中不断衍进、发展，但在题材内容方面不能逾越当时的等级制度。官式彩画在废除帝制时代以后又有了新的发展。

Brief Introduction to the Official Decorative Polychrome Paintings in the Dynasties of Ming and Qing (1368 — 1911)

by Wang Zhongjie

Growing up with the development of the Chinese architecture, the decorative polychrome painting on the surface of wood components is an essential part of traditional Chinese buildings. Like other elements in Chinese buildings, it had been evolving from simpleness to complexity, and from crudity to refinement. Buried in oblivion, early decorative polychrome paintings have not been found yet, leaving us a vague picture of it. Fortunately, a few rare and precious remains dating from the 7th to 14th century were discovered, and correspondingly, the government-compiled standards *Yingzao Fashi* (the Building Rules) were published in the Song Dynasty (960 — 1279), which offers us some important documentary and visual proofs. Later decorativeo polychrome paintings become more accessible than ever. There are plenty of authentic historical remains in dynasties of the Yuan (1279 — 1368), the Ming (1368 — 1644), and the Qing (1644 — 1911), together with another important government standards *Gongcheng Zuofa* (the Construction Method) issued during the Reign of Emperor Yongzheng (1723 — 1735). What is more, even today, veteran craftsmen in Beijing still follow a self-contained, conventional method in painting and renovating the traditional architectural decorative polychrome paintings. The above three factors would establish a substantial base for us to clarify and study the lore of the Ming-Qing official decorative polychrome paintings on both the cultural and technical aspects.

In following paragraphs, I would like to make a brief introduction to the Ming-Qing official decorative polychrome paintings.

The period of the Ming and Qing dynasties (especially the latter) was the most active and fruitful age for developing architectural decorative polychrome paintings in China. New variety kept emerging; themes were expanding broader; representation skills excelled the predecessor's; standards and orders became more rigid; hierarchical grades of the decorative polychrome paintings were inclined to be stricter and clearer. These achievements certainly owed a lot to the foundation laid in previous dynasties, yet certain historical circumstances fueled the development of decorative polychrome paintings as well, by providing both the cultural and physical conditions. For instance, the enrichment of colors (especially the steep rise of the annual output of gold foils) greatly promoted the splendid tendency in decorative polychrome paintings. Meanwhile, the rapid development on themes and techniques of the Qing decorative polychrome paintings could be also ascribed to other sources: the blending between the traditional Chinese culture and the minority cultures, the popularization of the literati drawings and the folk art, and the introduction of foreign cultures.

Generally speaking, the Ming-Qing decorative polychrome paintings may be classified into two categories: the official style and the local style. The former is a kind of formulized product made by official artisans. Under the direction of the Management of Building, products should match for certain hierarchical system or building material quota. Such category is usually applied in imperial buildings, aristocrats and ministers' mansions, imperially established temples, and government offices in the capital. The other category, the local one made by non-official artisans, however, is always applied in local government offices, local temples, and vernacular dwellings. It is more vivid, natural, and less rigid than the official one at formulation, yet still following the regulated hierarchical system. Of course, one category is different from another; meanwhile, both of them are mixing with each other in many ways. The local one once made its effort to imitate the official one, while local elements in the same time were largely absorbed into the official one as valuable supplements. Taking one with another, the official category is the senior and more orthodox one than any others in the Ming-Qing decorative polychrome paintings.

I. The Ming official decorative polychrome paintings

Since the survived Ming examples are quite rare, it is hard for us to get a full view on the decorative polychrome paintings in this period. It remains to explore that how many genres once existed for the Ming official decorative polychrome paintings. So far, only two genres were explored from material objects: one is the *xuanzi* (literally 'the spiral flower') pattern ('*xuanzi*' may be applied as a tentative term for describing the Ming official decorative polychrome paintings, since it remains to decide whether a more accurate name is available); the other one, being applied in buildings in imperial gardens, is with the dragon pattern in the central portion of painted beam and with the brocade pattern in the intermediate portion (such type looks quite similar to the official Suzhou style pattern in the middle period of the Qing dynasty). Among the historical remains today, the overwhelming majorities are of the *xuanzi* pattern, leaving only one single example for the second genre. Accordingly, following sentences will focus on the genre of *xuanzi* pattern and make relevant analyses on it.

The Ming official *xuanzi* pattern was evolved from its predecessor in the Yuan dynasty. The evolvement can be seen clearly from various historical remains. While the Yuan decorative polychrome paintings on wooden structure were diverse in composition, the Ming successors are shaped in regular arrangement. The focal point of evolvement is about the central portion and the intermediate portion of painted beam. The boundary of the central portion is always in '}' shape. In most cases, there are no detailed decorations but only plain colors painted in the central portion, namely the so-called 'plain central portion'. Such composition was quite common in those predecessor examples. As for the intermediate portion, the standard decorative element is the spiral flower, usually making-up a basic motif constituted with one whole flower and two half ones. The pattern of the spiral flower has been roughly formulized by that time; still we can find some traces where the paintings were sketched realistically. For example, in the central part of the spiral flower may be sketched a lotus throne, and above which added some seedpods or buds. Besides, while the spiral flower played a leading role then in decorating the intermediate portion, some other elements such as the *ruyi* head were also applied alternatively, in order to avoid the defect that the same motifs appear overly. Finally, the end portion of painted beam has been comparatively more formulized than other portions in the Ming official decorative polychrome paintings. That is to say, the end portion should always act as the ending of painted beam, regardless of their lengths.

The scope for employing the *xuanzi* pattern in the Ming dynasty was quite broader than that in the previous Yuan dynasty and the subsequent Qing dynasty. The Ming *xuanzi* pattern what we touch today is usually painted with colors on the surface of wooden components. In addition, the *xuanzi* pattern may also be made of other materials, for example, being fired in colored glaze decoration (e.g. colored glaze gates in Imperial Mausoleums of the Ming dynasty, Beijing, and in Wudang Mountain, Hubei), engraved with lines on brass components (e.g. brass roof in Wudang Mountain), or chiseled on the surface of stone components.

As I mentioned above, although those kinds of *xuanzi* pattern are made of disparate materials, scattering in spots at great distances from each other, however, nearly all of them are designed under the same standard. Thus it can be seen, the Ming official *xuanzi* pattern, while regularly formulized, was shaped by a full set of standards and orders, which were much more systematized than that in the past dynasties. In the Ming official decorative polychrome paintings, decorations on the surface of bracket sets *(dougong)* have evolved a manner of blue-green gradient to replace the old manner of painting detailed patterns on it. For decorative polychrome paintings on the ceilings, its composition has already been designed in the shape of square or round drums. As for the round drum shape, its detailed decorations are quite diverse, including the passionflower pattern, the Sanskrit scripts, and so on.

i. Color system and technique features of the Ming official decorative polychrome paintings

For the Ming official decorative polychrome paintings, its decorative pattern has some unique styles, and so do the color system and technique features. No full blank of background would be left for sketch drawings; further, red is seldom employed, with deep blue and green playing leading roles instead. Indeed, the impression that the Ming official decorative polychrome paintings look simple and elegant owes a lot to the well-chosen colors and refined techniques.

1. Color. During this period, all of the applied colors in decorative polychrome paintings are made from the indigenous mineral substances: mineral blue, mineral green, vermilion, etc. Bright but not raffish as they appear, the colors are quite similar to those in traditional Chinese paintings. Composed by these colors, the Ming decorative polychrome paintings look like some kinds of three-dimensional ornamental paintings in vivid blue and green. In fact, during the past decades, the academy tried several times in filling the plain wooden structure of old buildings with the Ming decorative polychrome paintings, but always received little satisfaction on hue aspect. Why? One answer in good reason is that the applied colors are all modern ones such as Paris green and ultramarine; they are brighter yet less decent than the mineral colors. Thus it can be seen, the elegance of the Ming official decorative polychrome paintings is indeed indivisible from the well-chosen colors.

2. Hue. The hue aspect is characterized by its collocation of warm hues and cool hues. For instance, on the surface of painted beam, red is embellished on the blue-green background of the central lotus throne. Such method effectively stresses the 'lotus' motif and breaks the rigid painting discipline. The same method is also suitable for applying gold foils. To avoid make the decorative polychrome paintings too ornamental, the gold foils always, instead of being employed everywhere, focus on certain prominent spots, such as stamens, pistils, or rhombus corners. Further, they are rarely arranged symmetrically lest the hue composition appears monotonically.

3. Manner. The main technique features of the Ming official *xuanzi* pattern could be concluded as: regardless of their hierarchical grades, all of the decorative polychrome paintings are in the manner of color gradient, from lighter to darker, darkening level by level moderately. Concerning the material objects survived today, the lightest color to draw an outline is not white but light blue or light green. In this way, the whole painting appears gentle, since the highly contrasted whiteness is out of use. Besides painted beams, other structural components are also in the manner of color gradient, thus shaping the unique Ming style: simple, elegant, and gentle.

ii. Hierarchical grades of the Ming official *xuanzi* pattern

Our lore on hierarchical grades of the Ming official *xuanzi* pattern so far is severely restricted by the lack of survived examples and detailed literature records. Among the survived Ming decorative polychrome paintings, those on the highest hierarchical grade are for palaces in which the empresses and imperial concubines lived; and the second highest, for certain halls in the corner courtyards of the imperial palace, and some sacred buildings. However, none of the Ming decorative polychrome paintings have been found in buildings for the use of the emperors, in government offices, or in vernacular dwellings, hence much increasing the difficulty for further research. It can only make a partial inference now from those existed material objects, that the *xuanzi* pattern was the dominant type of decorative polychrome paintings applied in the Ming dynasty. While still invisible as they are for the decorative polychrome paintings in Fengtian Palace (the palace witnessing the emperors' enthronements and governing), we could somehow take certain examples form sacred buildings as the available reference. Besides, the decorative polychrome paintings in government offices may be traced from the image in temples. In short, the hierarchical grades of the Ming official *xuanzi* pattern may probably hinge on how delicate the paintings are, and how many gold foils are employed. In this way, three grades could be set down in general:

1. **Gold lines and largely gold-pointing pattern.** The central portion of painted beam is probably the 'plain central portion' without detailed decorations, or would rather, applying the dragon pattern instead. The main contour lines are totally made in the manner of embossing and gold foil gilding, and so do the flower centers and the rhombus corners. The ceiling is also framed with gold lines as what the beam does, and makes partially gold pointing on detailed decorative spots. And the bracket sets are painted in the manner of blue-green gradient. To conclude, the above manners are for the highest hierarchical grade of Ming decorative polychrome paintings, which are suitable for main halls in the imperial palace, important shrines, and temples.

2. **Black lines and gold-pointing pattern.** The central portion of beam is not painted with detailed decorations. The main contour lines are totally painted in black ink, while the flower centers and the rhombus corners are embellished with gold pointing. (The free options for applying gold pointing in different spots and amounts may block us from defining it as a precise type of gold pointing. Rather, we would offer it a general term 'the gold-pointing pattern'.) The ceiling is also decorated by gold pointing, but in less amounts. The bracket sets are painted in the same manner as what the 'gold lines and largely gold-pointing pattern' has done. To conclude, the above manners are for the in-between hierarchical grade, and consequently suitable for wider range, such as halls in the imperial palace where empresses and imperial concubines lived, some other relatively important imperial buildings, or main halls in various temples.

3. **Non-gold-pointing pattern.** Such hierarchical grade is suitable for ordinary constructions.

The Ming literature once recorded the term 'Suzhou style pattern'; however, the absence of relevant material objects keeps us out from learning their concrete images.

Quite rare today, the survived examples of the Ming local decorative polychrome paintings could be roughly divided into two types. One is the *xuanzi* pattern in buildings of the Northern temples. Its decorative composition and color system are almost the same as those of official styles. Less restrictive and more flexible, examples in the local type suggest that they are in fact the imitations of the official one.

Another type is the decorative polychrome paintings in the Southern dwelling houses. Its basic decorative element is the 'cloth-wrapper'. Comparing with the official style, its detailed decoration and color system are even more delicate and elegant, which reflects the charming Southern style.

II. The Qing official decorative polychrome paintings

Like other art works in China, it was by little chance that the decorative polychrome paintings would have taken a sudden change during the dynastic changes in the 1640s. Rather, there should be a gradually evolving process. During the period of the Qing dynasty, the official decorative polychrome paintings were greatly promoted by the enrichment of colors, the unprecedented flourishing of the fine arts, and the splendor tendency in dynast lives.

The genres of the Qing official decorative polychrome paintings are much more diverse than those in the Ming dynasty. According to different frame compositions and themes, at least five decorative genres could be listed, namely: the *hexi* (combined dragon seals) pattern, the *xuanzi* (spiral flower) pattern, the Suzhou style pattern, the lucky grass pattern, and the full covering pattern.

i. *Hexi* (combined dragon seals) pattern

The *hexi* pattern, while on the highest hierarchical grade of the Qing decorative polychrome paintings, actually took shape much later than other genres of decorative polychrome paintings. It didn't emerge until the middle period of the Ming dynasty, and finally shaped up during the dynasty change in the mid-17th century. It stands to basic compositions of the *xuanzi* pattern since being derived from it. Indeed, both the *hexi* pattern and the Ming *xuanzi* pattern are composed of three same portions: the end portion, the intermediate-central-intermediate portion, and the end portion again. What distinguishes the *hexi* pattern from its predecessor is that, three horizontal frames in lotus shapes have replaced the previous diverged 60-degree-angle strips that defined the intermediate portion of painted beam, and the original spiral flowers have been taken out. Another point about the decoration of hexi pattern is that, it is not the brocade pattern and flowers, but the dragon pattern, the phoenix pattern, the passionflower pattern, and the lucky grass pattern being painted on the central portion, the intermediate portion, and the box of beams, and on components such as flat ties and cushion boards. Concerning the color system, the hexi pattern is distinguished from any other genres. For its main contour lines, instead of being painted in black live, are totally made in the manner of embossing and gold foil gilding. And so do most of the detailed decorations.

At different themes of the detailed decorations, the *hexi* pattern could be classified into following five sorts.

1. **Dragons pattern.** None of other motifs but the dragons pattern is painted on the central portion, the intermediate portion, and the box of wooden beams, and on components such as flat ties, cushion boards, and capitals. The academy once called this sort 'total dragons pattern' or 'five dragons pattern': namely the dragons pattern on purlins, on cushion boards, on flat ties, on architraves, and on capitals. However, it cannot be regarded as an accurate account, because it has excluded some other components being decorated with the dragon pattern. Therefore, since the basic motif in such sort is the dragon pattern, I would rather call it 'dragons pattern'.

2. **Phoenixes pattern.** Just as its name implies, none of other motifs but the phoenix pattern is painted on the central portion, the intermediate portion, and the box of wooden beams, and on components such as flat ties, cushion boards, and capitals.

3. **Dragons and phoenixes pattern.** The dragon pattern and the phoenix pattern are painted alternatively in this sort.

4. **Dragons and phoenixes core pattern.** The dragon pattern and the phoenix pattern are painted on the central portion and the box of beams; the passionflower pattern and the ganoderma lucidum pattern are painted on the intermediate portion of beams.

5. **Dragons and grass pattern.** The dragon pattern and the lucky grass pattern are painted alternatively on the central portion, the intermediate portion, and the box of wooden beams, and on components such as flat ties and cushion boards.

Besides the above ones, the hexi pattern applied in the Buddhist temples are often painted with religious decorations like Sanskrit scripts, pagodas, and lotus thrones.

While the wooden beam is the focus of decoration in the *hexi* pattern, other relevant components could also be very delicate. In the high-grade *hexi* pattern, the rafters and roof boarding may be largely embroidered, and the bottom surface of hip rafters could also be painted in the dragon pattern. Regarding the bracket sets, no matter which sort of *hexi* pattern is applied, all the contour lines should be edged by gold foils.

Judging by its color system and relevant technique, the *hexi* pattern is a relatively straightforward kind of decorative polychrome paintings, since its decoration is almost made in the manner of embossing and gold foil gilding. The only exceptions are the diverged box corners and the cloud patterns in the intermediate portion of painted beam, where may employ the manner of 'pressing back (*zatui*)', painting the background in a monochromatic color. In the early *hexi* pattern, a light blue or light green line would be added on one side of the main contour lines. Yet in the late *hexi* pattern, a substituted blur color is added along the main contour lines, and a white line replaces the previous light blue or light green one, hence demonstrating a varied flavor. One subtle point of embellishment in the *hexi* pattern is the hue contract between gold lines and the blue-green background. The dark grounding underlines the dazzling gold decorations, thus making a splendid atmosphere of the whole building. In normal conditions, the *hexi* pattern employs only one kind of gold foil, but as for the high-grade *hexi* pattern, gold foils in both dark and light colors (namely the 'red gold foil' and the 'yellow gold foil') could be employed in order to increase profound feelings.

Although all sorts of the hexi pattern are on a high hierarchical level among the Qing decorative polychrome paintings, each of them is in fact oriented to a specific grade. The 'dragons pattern' is in the first grade of the hexi pattern, only suitable for palaces in use for the emperors' enthronements and governing, and for the main halls in important temples. Inferior to the 'dragons pattern', the 'dragons and phoenixes pattern' and the 'dragons and phoenixes core pattern' are in the second grade of the hexi pattern. They are suitable for imperial sleeping palaces and for the main halls in temples that hold sacred ceremonies for the Heaven, for example, Qinian Palace (the palace for praying for a year of abundance) in the Temple of Heaven, Beijing. The 'phoenixes pattern' has a less wide employing range, merely suitable for the empresses' sleeping palaces and for the main halls in temples that hold sacred ceremonies for the Earth, for example, Huangqi Palace (the imperial palace offering sacrifice to god of the earth) in the Temple of Earth, Beijing. It should also be classified in the second grade. Finally, the 'dragons and grass pattern' is in the lowest grade of hexi pattern, suitable for main gates in the imperial palace, side halls along the central axis, and main halls in important temples.

ii. *xuanzi* (spiral flower) pattern

The Qing *xuanzi* pattern was derived from the Ming predecessor. Being a prominent genre in the Qing official decorative polychrome paintings, the *xuanzi* pattern is of various sorts and in wide use. In general, eight sorts can be classified according to the decorative composition, the color system, and the technique manner, namely:

1. Totally gold foil gilding pattern;
2. Gold foil gilding and blue-green gradient pattern;
3. Black lines and blue-green gradient pattern;
4. Gold lines and largely gold-pointing pattern;
5. Black lines and largely gold-pointing pattern;
6. Spot gold-pointing pattern;
7. Elegant black pattern;
8. Mineral yellow pattern.

The common point of the above all sorts of *xuanzi* pattern are about their general composition. If the beam is long enough, a box may frame next to the two end portions of painted beam, with sub-end portions setting apart inside the box. Next, a central portion of the beam is marked at about one third of its total length. Between the central portion and the end portion, then, is what we call the 'intermediate portion'. It may be filled with circle flowers (spiral flowers), which are composed by multi-layer spiral petals. The most common combination form of flowers is that with one whole flower and two half ones (Such combined motif is usually called 'one whole flower and two halves'). According to the varied length of beams, an adjustment should be made for the intermediate portion, and the proper combination form of spiral flowers should also be planned. The motif of 'one whole flower and two halves' can be repeated several times if the beam is long enough; in case the beam is short, one spiral flower should rather be lapped over another or cut itself into portions. Further, the naming 'spiral flower *(xuanzi)*' is derived from the periphery spiral-shaped petals of the circle flowers. Therefore, the above eight sorts of decorative polychrome paintings could all be called the '*xuanzi* pattern' for their common decorative points: the spiral flower.

Regardless of different grades of the *xuanzi* pattern, the spiral flower is immutable as the basic decorative element in the intermediate portion of painted beam. On the contrary, detailed decorations in the central portion and in the box could be varied along with the hierarchical grades. For instance, from higher grades to lower ones, the central portion may be painted respectively in the dragon pattern, the dragon-and-phoenix pattern, the phoenix pattern, the brocade pattern, the grass-shaped dragon pattern, the scroll grass pattern, and the flowers, etc. In some of the low-grade *xuanzi* patterns, the central portion may even be filled nothing but a plain black grounding, without any detailed patterns on it. As for the 'box', it is a square between the two end portions of painted beam, with its length approximately the same as the height of architrave. For the high-grade *xuanzi* patterns, boxes are framed by a nearly circular contour that constituted with eight arcs, and filled with decorations such as the dragon pattern, the passionflower pattern, or some rare animals. Such composition is recognized as the 'vivid box'. As for boxes in the low-grade *xuanzi* patterns, the geometric-formed contour without a circular frame could be composed of four gardenia patterns in petal shape, which we call the 'rigid box'.

Below I will briefly introduce the specific manners on employing each sort of the *xuanzi* pattern, according to their varied decorative compositions and color systems.

1. **Totally gold foil gilding pattern.** None of detailed decorations is set in the central portion of beam. Only the gardenia pattern may be embellished on the box. Without painting colors, all decorations are made in the manner of embossing and totally gold foil gilding.

2. **Gold foil gilding and blue-green gradient pattern.** The dragon pattern, the grass-shaped dragon pattern, the phoenix pattern and so on are painted on the central portion and the box of beam. Both the main contour lines and the detailed decorations employ the manner of embossing and then totally gold foil gilding. The main contour lines and spiral flowers are all painted in the manner of blue-green gradient.

3. **Black lines and blue-green gradient pattern.** Decorations in the central portion and the box of beam are made in the same manner as in the 'gold foil gilding and blue-green gradient pattern'. Options of employing the plain central portion and the rigid box are also available. Main contour lines, spiral centers, rhombus corners, and gardenia centers in the intermediate portion, as well as decorations in the central portion and the box of beam are all made in the manner of embossing and gold foil gilding. The edges of spiral flowers and gardenias are drawn in black ink. The main contour lines and the whole spiral flower are all painted in the manner of blue-green gradient.

4. **Gold lines and largely gold-pointing pattern.** Decorations in the central portion of beam are mainly painted in the dragon pattern and the brocade pattern, which in professional terms called the 'dragon and brocade central portion'. Yet it is also allowable to solely employ the dragon pattern. Decorations in the box are mostly painted in combination of the dragon pattern and the passionflower pattern. Main contour lines, spiral centers, rhombus corners, and gardenia centers in the intermediate portion, part of the dragon pattern and the brocade pattern in the central portion, as well as the dragon pattern and the passionflower pattern in the box are all made in the manner of embossing and gold foil gilding. The main contour lines are all painted in the manner of blue-green gradient. The edges of spiral flowers and gardenias are painted in black ink on the blue-green grounding, and then added with white chalk traces along the inside boundary lines.

5. **Black lines and largely gold-pointing pattern.** In most cases, no detailed decorations would be arranged in the central portion of beam, yet individual examples with the dragon pattern and the brocade pattern are also available. The box is mainly the 'rigid box'. Spiral centers, rhombus corners, and gardenia centers are made in the manner of embossing and gold foil gilding. (The painted dragon pattern, the brocade pattern, and the passionflower pattern in the central portion of beam may be gilded by gold foils totally or partially.) Main contour lines, spiral flowers, and gardenias are all painted in black ink, added with white chalk traces along the boundary lines.

6. **Spot gold-pointing pattern.** In most cases, the central portion of beam is filled with the grass-shaped dragon pattern and the black-leaf and green-grounding flowers. The combination of the two decorations is called in professional term 'grass-shaped dragon and flowers central portion'. Yet the manner of 'plain central portion' with no detailed decorations is also available, whileplus the box is usually the 'rigid box', and the gold foil gilding is focused on spiral centers and gardenia centers. The main contour lines are painted

in black ink. And the contour lines of spiral flowers (including rhombus corners) are simply painted in black ink on the blue—green grounding, and then added with white chalk traces along the boundary lines.

7. Elegant black pattern. The decorations are all the same as in the 'spot gold-pointing pattern', except for that the gold foil gilding is out of use. The detailed manner is also the same as in the 'spot gold-pointing pattern'.

8. Mineral yellow pattern. The decorations are basically the same as in the low-grade *xuanzi* patterns, usually in the manner of 'plain central portion' and 'rigid box'. For the color system, this sort of decorative polychrome paintings is extremely differed from other sorts of *xuanzi* pattern. Its dominant colors in use are not blue or green, but with all the grounding painted in mineral yellow, and main contour lines, spiral flowers, and gardenias are all in black ink. On the assumed blue grounding, only a blur, light blue line may be drawn along the boundary lines, added with white chalk traces above. The same method would be applied for the assumed green grounding, that only a light green line is drawn. Usually, the gold foil gilding is out of use for the 'mineral yellow pattern'.

For the *xuanzi* pattern, the decorations on components other than the wooden structural beam should match the ones on the latter. The highest four grades of the *xuanzi* pattern, for instance, the 'gold lines and largely gold-pointing pattern', should match with following decorations: dragon-eye pearls on the eave rafter ends, Swastika patterns of gold foils on the flying rafter ends, gold edges on the bracket sets, gold edges and blurred gold grounding on the hip rafters and several kinds of beam ends, gold contour lines on the sparrow braces *(queti)* and carven boards, and the manner of black 'pressing back' on the woodcarving work. As for mid-grade *xuanzi* patterns, such as the 'black lines and largely gold-pointing pattern' and the 'spot gold-pointing pattern', rules of decorations on components are not so strict. On the prominent places like the hip rafters and beam ends, both the manners of 'gold edges and black grounding' and 'black edges and black grounding' are allowable. Also, both the manners of rendering whitening and 'pressing back' are allowable in the woodcarving work of sparrow braces and carven boards. However, contour line of the bracket sets in mid-grade are all in black.

For the 'elegant black pattern' in the lowest grade of *xuanzi* pattern, decorations are set as below: black Swastika patterns on the flying rafter ends, tiger-eye pearls on the eave rafter ends, 'black edges and black grounding' on the bracket sets and the hip rafters, yellow contour lines on the sparrow braces and carven boards, and the manner of rendering whitening on the woodcarving work.

As the second genre of the Qing official decorative polychrome paintings, the *xuanzi* pattern in practical uses could be employed approximately under four situations.

First, the *xuanzi* pattern is often employed in secondary constructions in the imperial palace and imperial gardens, like ordinary palaces, corridor houses, and gatehouses. The most usual sorts in use are the 'gold lines and largely gold-pointing pattern' and the 'black lines and largely gold-pointing pattern'. Certain important palaces may also employ the 'gold foil gilding and blue-green gradient pattern' or the 'black lines and blue-green gradient pattern'. Low-grade constructions such as the gatehouses are usually decorated in the 'spot gold-pointing pattern' or the 'elegant black pattern'.

Second, the high-grade *xuanzi* pattern is employed in halls inside or outside the imperial palace that hold sacred ceremonies for the ancestors, (e.g. Fengxian Palace and the imperial Great Temple) and in main halls in imperial mausoleums. For example, the 'totally gold foil gilding pattern' is employed in Fengxian Palace (the palace holding sacred ceremonies for the ancestors) in the Forbidden City, and the 'black lines and blue-green gradient pattern' is employed in the main palace of the Great Temple. The latter sort could also be employed in main halls of the Qing imperial mausoleums.

Third, the *xuanzi* pattern is also employed in secondary buildings in important shrines and temples, and in ordinary temples and palag.

Fourth, the 'mineral yello~Ppattern' is one kind of decorative polychrome paintings for special uses, mainly empteyed in kitchens that prepare sacrificial offerings, such as the sacred kitchens and storehouses in imperial mausoleums. But there also exist a few exceptions, for example, the 'mineral yellow pattern' once was employed in Yuegu Pavilion (the storied building for collecting precious calligraphy copies) in Beihai imperial garden.

iii. Suzhou style pattern

The Suzhou style pattern is a kind of decorative polychrome paintings that for decorating buildings in the landscape gardens. It originated in Suzhou area in the Southland, and finally turned into an important type of official decorative polychrome paintings after it came into the imperial court in the Northland. By far, most of the decorative polychrome paintings in early official Suzhou style pattern we see today are the survived examples during the Emperor Qianlong's reign (1736 – 1796). From general compositions to detailed decorations, it has become totally official and northern, no longer Äith% ⊠ aces of the original Suzhou style. Although the naming 'Suzhou style' continues to use, it has in fact evolved itself as a type of decorative polychrome paintings with unique style. Generally, the Qing official Suzhou style pattern could be classified in two sorts: those in the early and middle period, and those in the late period.

1. Official Suzhou style pattern in the early and middle period. According to its composition forms, the Suzhou style pattern in this period may be divided into three types approximately, namely the central portion type, the cloth wrapper type, and the full covering type.

The main contour lines of the 'central portion type' are similar to those of the *xuanzi* pattern. That is to say, it copies the main contour lines of the *xuanzi* pattern, while variations mainly focus on the intermediate portion, where the spiral flowers have been replaced by decorations such as the brocade pattern, the circle flowers, the clip frame, and the brocade frame. The central portion of painted beam remains almost unaltered, employing the dragon pattern, the phoenix pattern, and the passionflower pattern as before, or at most, adding a few ancient-style paintings or sketches.

The divergence in composition between the 'cloth wrapper type' and 'central portion type' is that the former has removed the central portion, and then designs a picture frame in semicircle shape that covers the central portions of beams, cushion boards, and architraves. Such a picture frame, resembling a sagged circle cloth, is accordingly called 'cloth wrapper' from its image. Consequently, this composition form is called as the 'cloth wrapper type'. At its intermediate portion, it is almost the same as that of the 'central portion type'. As for the 'cloth wrapper', it is always filled with blessing patterns that symbolize nice wishes, for example, the patterns for wishing for the health and happiness.

As for the 'full covering type', only the end portion of painted beam remains unaltered. All of the central portion, the intermediate portion, the cloth wrapper, and the pool frame have been removed. No frame is set, hence forming an open appearance. Or at most, a pair of clip frames is painted on the end portion of beam, then adding the scroll grass pattern, the bat-and-jade pattern, or the black-leaf flowers above. Viewing from the detailed decorations, motifs in the Suzhou style pattern during this period were nearly the same as the *hexi* or *xuanzi* patterns: mainly the dragon pattern and the phoenix pattern, sometimes the certain blessing patterns, and occasionally sketches. Such composition motifs led to similar technique manners to those of the *hexi* or *xuanzi* patterns: blue-green grounding, embossing, and gold foil gilding. What has been differed is that the manner of 'pressing back' became more important than ever, and there added a few sketches for arranging.

The official Suzhou style pattern during this period had little *joie de vivre*, and it was also far from vividness. It was a totally regular, serious kind of imperial decorative polychrome paintings. By far, no examples in the dignitary dwellings in the early and middle period have been discovered yet. Therefore, it is too difficult to know the real cases of such decorative polychrome paintings in that period.

2. Official Suzhou style pattern in the late period. On the composition aspect, the Suzhou style pattern in this period could also be divided into the same three types, namely, the central portion type, the cloth wrapper type, and the full covering type. However, the changes in detailed decorations are quite obvious. For the 'cloth wrapper', the solely employed cloud-shaped frame has replaced the coexisted lace frame and cloud-shaped frame. For the end portion of painted beam, decorations of the Swastika pattern and the fret pattern take the place of the original plain end portion. And for the intermediate portion, the brocade pattern has no longer been arranged on it. However, what changed the most in the 'cloth wrapper type' (including the 'central portion type') should be those drawings inside the 'cloth wrapper', the central portion, and the pool frame. The previous patterns have evolved into various sketches. In fact, the sketches as well as the 'brocade frame' constitute the dominant image of the late Suzhou style pattern.

The sketches involve portraits in certain narratives, realistic paintings about flowers and birds, ink paintings with idyllic feelings on describing buildings and landscape, etc. The implication in sketches is for wishing for fortune, though it is less straightforward than that in blessing patterns. While the 'central portion type' in the late period appears almost the same as the 'cloth wrapper type', the composition of the late 'full covering type' has changed little from the early and middle period. The flowing cloud pattern and the creeping flowers emerge as the new decorative elements, and the main theme of the 'full covering type' has been arranged as the flowing cloud pattern and the black-leaf flowers.

For the Suzhou style pattern, decorations on components other than the wooden structural beam should also match the ones on the latter. For instance, when the manner such as gold lines, gold pointing, or gold foil gilding is employed on structural beams, the same manner should be employed on the eave rafter ends and the flying rafter ends. If the building is a wooden-structural one with bracket sets, gold contour lines will then be painted on the edges of the bracket sets. Once the manner of yellow or black lines is employed for the lower-grade structural beams, it should correspondingly reduce to the manner of 'yellow grounding and yellow edges' or 'black grounding and black edges' for employing on the eave rafter ends and the flying rafter ends.

It is somehow difficult to make a clear division on grades of the Suzhou style pattern, because it is differed from the *hexi* or *xuanzi* pattern. The latter two, ruled by strict hierarchical grades, are easy to make out every sort according to its detailed decoration grade. What's more, the order of grades cannot even be inverted, and the rules are set down so carefully that even the individual spots or different amounts for employing the gold pointing have been determined. On the contrary, either the decorations or technique manners of the Suzhou style pattern are quite unrestricted, especially in the late period of the Qing dynasty. From examples we see today, the Suzhou style pattern in the early and middle period may be divided approximately into two grades: (1) the high-grade Suzhou style decorative polychrome paintings are mainly painted in gold lines. Detailed decorations such as the dragon pattern, the phoenix pattern, the circle flowers, the brocade pattern, and the clip frame are all made in the manner of gold pointing or gold foil gilding 'pressing back'; (2) the low-grade Suzhou style decorative polychrome paintings are mainly painted in black lines. Detailed decorations such as the clip frame, the bat-and-jade pattern, and the scroll grass pattern are almost made in the manner of black lines 'pressing back', yet sometimes also embellished partially by the gold pointing.

The Suzhou style pattern in the late period may be divided approximately into three grades: (1) the high-grade Suzhou style decorative polychrome paintings are mainly painted in gold lines. Detailed decorations in the end portion and the clip frame of painted beam are always made in the manner of gold foil gilding 'pressing back'. Sketches inside the central portion, the 'cloth wrapper', the brocade frame, and the pool frame primarily concern on buildings, landscape, flowers, and birds, etc.; (2) the mid-grade Suzhou style decorative polychrome paintings are mainly painted in gold lines. Detailed decorations in the end portion and in the clip frame of painted beam are usually made in the manner of gold pointing or black lines 'pressing back'. Sketches inside the central portion and the 'cloth wrapper' correspondingly reduce in amounts; (3) the low-grade Suzhou style decorative polychrome paintings are mainly painted in yellow lines. The end portion is always the 'plain end portion', while detailed decorations in the clip frame of painted beam are totally made in the manner of black lines 'pressing back'. Sketches inside the central portion and the cloth wrapper reduce even further.

The hierarchical grade system of the Suzhou style pattern is undoubtedly hard to determine. Generally speaking, high grades are employed in primary constructions, and low grades are employed in secondary constructions. However, at least one point is for sure, that the dragon pattern and the phoenix pattern, which signify the imperial power, are disallowed outside the range of imperial buildings.

iv. Lucky grass pattern

In the early and middle period of the Qing dynasty, there existed another kind of official decorative polychrome paintings: the pearl and lucky grass pattern, which is with quite unique composition and flaming hue. Referring its composition, the two end portions of beam are first defined, and then, the central portion of beam is filled with a huge flower circle. Three pearls are placed in the center of flower, and the encircled scroll grass pattern is painted around the pearls. The flower circle spreads out on both sides from the bottom of component, just like a huge, reversed 'cloth wrapper'. On the upper side inside the two end portions of beam, a combination of the scroll grass pattern is employed. The purlin is thin yet long, so the employed decorative pattern is composed of only three pearls in the center plus scroll grass pattern along the two sides. Next to the end portions of beam, two groups of scroll grass are painted respectively. (Such composition is quite similar to the shape of male-female grass during the middle and late period.) Concerning the color system of the lucky grass pattern, its grounding is totally painted bright red; the scroll grass pattern is painted in the manner of 'gold petal and blue-green grounding pressing back'; the outline border of the three pearls is made with the manner of embossing and gold foil gilding; the pearls themselves are painted dark blue and green. In this way, the complementary hue appears simple, straightforward, and quite flaming.

The lucky grass pattern was an independent genre in the early period of the Qing dynasty, usually employed in the gates of the imperial palace and in the imperial Mausoleums. Around the middle period, it had gradually blended with other types of decorative polychrome paintings, no longer emerging as an independent genre from then on.

v. Full covering pattern

The full covering pattern referred here is not the above-mentioned 'full covering type' of the Suzhou style pattern. It is a kind of decorative polychrome paintings applied on the total surface of certain buildings(even including the eave edgings, rafter connecting boards, upper wooden structures, lower wooden structures, etc.). It may probably come into being in the late period of the Qing dynasty. According to the decorations and color system, three technique manners could be concluded about the full covering pattern.

1. The mottled bamboo patterns (or in the artisan's word, the 'mottled bamboo base') are painted on the surface of all components. They are shaped in the bamboo stake patterns that are parallel to the orientation of components, looking as if each component has been composed of bamboo stakes. As for the structural beam, the two end portions are composed by thin bamboo stake patterns, and the central portion is filled with blessing motifs that composed of bamboo pattern combinations. The bamboos patterns are respectively painted into different forms of old bamboos or young ones, while speckles are painted on their nodes, so as to dress up the bamboo patterns like natural mottled ones. Elegant as it is, this sort of decorative polychrome paintings is to show the beautiful, natural scenery.

2. All components are painted green or primrose yellow as the grounding color. And next, trailing plants such as the creeping peony or the wisterias are depicted above. It is to create an environment where various flowers are all in full bloom.

3. All the upper wooden components are painted dark blue as the grounding color. And next, the polychrome flowing cloud patterns are drawn above.

The full covering pattern appears novel and interesting, yet it was just used in limited areas. In fact, the mottled bamboo pattern could somehow be employed occasionally, whereas the technique manners of other two sorts have been lost.

The Qing official decorative polychrome paintings are one of the main points reflecting the hierarchical grades of ancient official buildings. Kept evolving and developing, anyhow, they were impermissible to break their hierarchical grades by that time. It was until the downfall of the monarchy era in 1911 that a much broader perspective is presented for the new development of decorative polychrome paintings.

(张宇 翻译 / Translated by ZHang Yu)

15 明代彩画

北京智化寺明代金线旋子彩画　北京文物整理委员会

北京智化寺明代金线旋子彩画　北京文物整理委员会

17　明代彩画

北京智化寺万佛阁下层梵文天花　北京文物整理委员会

明代彩画 18

北京东四牌楼清真寺礼拜殿内檐包袱彩画　北京文物整理委员会

19　明代彩画

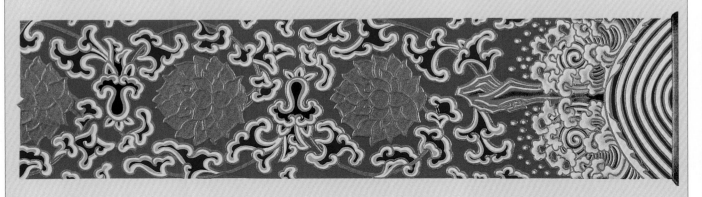

北京东四牌楼清真寺礼拜殿内檐柱彩画　北京文物整理委员会

北京磨石口法海寺山门外檐明代雅五墨彩画 杜恒昌

21　明代彩画

北京磨石口口法海寺山门脊部明代雅五墨彩画　高成良

明代墨线点金旋子彩画 杜恒昌

和玺彩画简述

蒋广全

和玺彩画是清代最高等级的一类彩画，装饰于清廷的宫殿、寝宫、坛庙、敕建庙宇等重要建筑。该类彩画出现于明末清初时期，系由旋子彩画的基本构图框架演变而来。其大木彩画纹饰仍保持了旋子彩画设有箍头、找头、方心"三段式"的构图格局，但其主体轮廓大线则以"ᔕ"形斜线，构成方心、岔口线、皮条线、圭线光等线造型为突出特征。清早期和玺大线的斜线画法用弧线表现，中晚期和玺大线斜线演变为用直线表现。和玺彩画的细部主题纹饰主要用象征皇权的龙凤纹，其次还有西蕃莲纹、吉祥草纹及仅用于重要佛教庙宇的梵纹等纹饰。

按和玺彩画细部主题纹饰的不同，大体可分为如下几种等级。

1. 龙和玺。梁枋大木中的方心、找头、盒子及平板枋、垫板、柱头等构件主要绘以龙纹。

2. 龙凤和玺。梁枋大木中的方心、找头、盒子及平板枋、垫板、柱头等构件主要绘以龙与凤相匹配的纹饰。

3. 龙凤方心西蕃莲灵芝找头和玺。梁枋大木中的方心及盒子绘以龙纹与凤纹，找头分别绘以西蕃莲及灵芝纹。

4. 凤和玺。梁枋大木的方心、找头、盒子及平板枋等部位主要绘以凤纹。

5. 龙草和玺。梁枋大木的方心、找头、盒子及平板枋、垫板等构件均绘以龙纹与法轮吉祥草纹，相互排列组合而成。

6. 梵纹龙和玺。梁枋大木的方心、找头、盒子及平板枋等构件均绘以梵纹与龙纹。

按不同的和玺品种装饰建筑有严明的等级制度。龙和玺为第一等，只适于皇帝登基、理政、居住的殿宇及重要坛庙；龙凤和玺、龙凤方心西蕃莲灵芝找头和玺、凤和玺为第二等，其中的前两种和玺适于帝后寝宫及祭天坛庙，后种凤和玺适于皇后寝宫及祭祀后土神坛的主要殿宇；龙草和玺、梵纹龙和玺为第三等，其中的龙草和玺适于皇宫的重要宫门及主轴线上的配殿及重要的寺庙殿堂，梵纹龙和玺仅适于敕建藏传佛教庙宇的主要建筑。

这类彩画的设色工艺特点是绝大部分纹饰都采用沥粉贴金做法（高等级的施用两色金手法，次级的施用一色金手法），盒子、岔角、方心、找头等部位的某些细部花纹，采用攒退等工艺做法。其青绿大色的分配，采取青绿相间的方法，且均为平涂。早中期和玺在主体轮廓大线旁只饰以白色线。晚期的和玺于大线旁绘以白色大粉及晕色，用以体现色彩的层次韵味。这类彩画用金量大，体现了皇家建筑豪华、庄重的气派以及金碧辉煌的装饰效果。

Hexi (Combined Dragonseals) Pattern

The *hexi* (combined dragonseals) pattern was of the highest rank in the Qing Dynasty and used in the important buildings such as palaces, emperors' dedrooms, temples, altars and temples built under the emperors' order. Such a type of drawings, first appearing in the end of the Ming Dynasty and the beginning of the Qing Dynasty, grew out of the basic framework of composition of the *xuanzi* (spiral flower) pattern. The theme patterns of its detail are the dragon pattern and the phoenix pattern which symbolize the imperial power. According to different patterns of the detail the *hexi* pattern can be classified into the following patterns:

1. Dragons pattern, the first grade
2. Dragons and phoenixes pattern, the second grade
3. Dragons and phoenixes core, the second grade
4. Dragons and grass, the third grade
5. Phoenixes pattern, the second grade
6. Sanskrit pattern, the third grade

25　和玺彩画

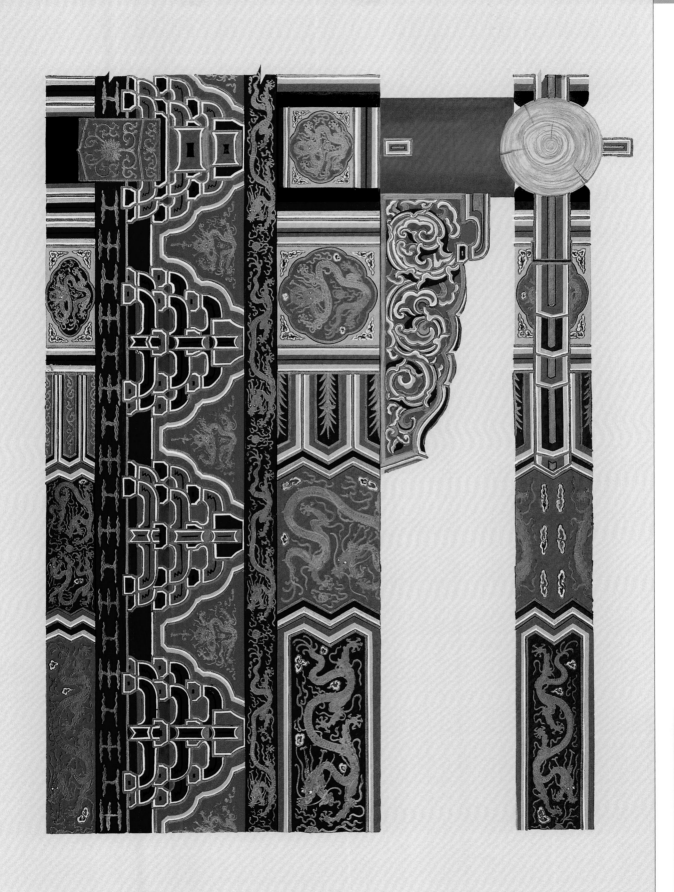

龙和玺彩画　杜恒昌　高成良

金琢墨龙和玺彩画 杜恒昌

龙凤和玺彩画 边楷一 杜恒昌

29　和玺彩画

凤和玺彩画　蒋广全

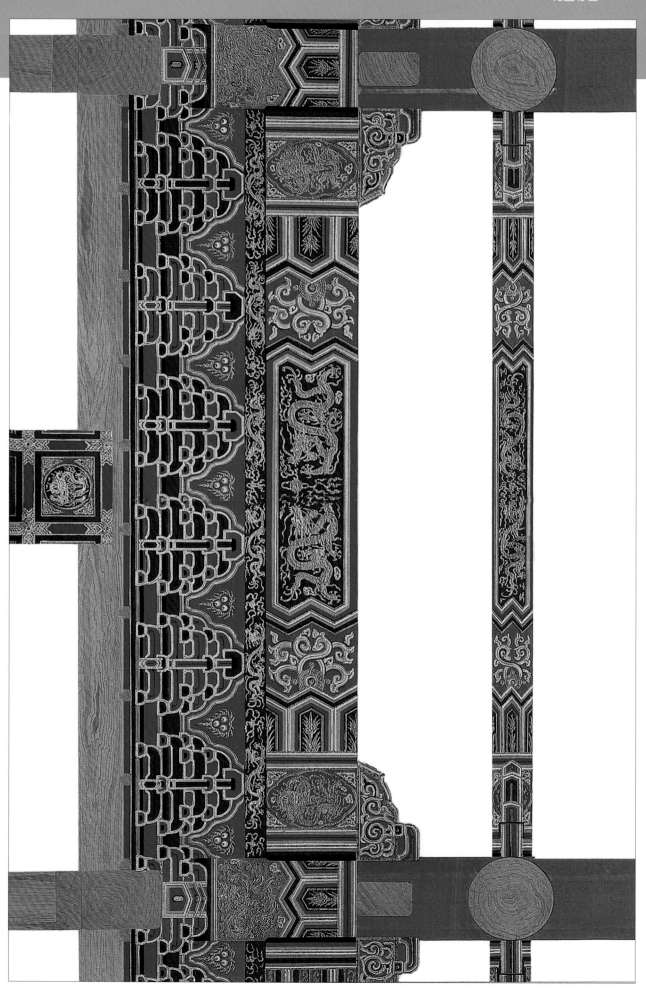

龙草和玺彩画　蒋广全

内檐梁架龙草和玺彩画 高成良

龙草和玺彩画　张秀芬

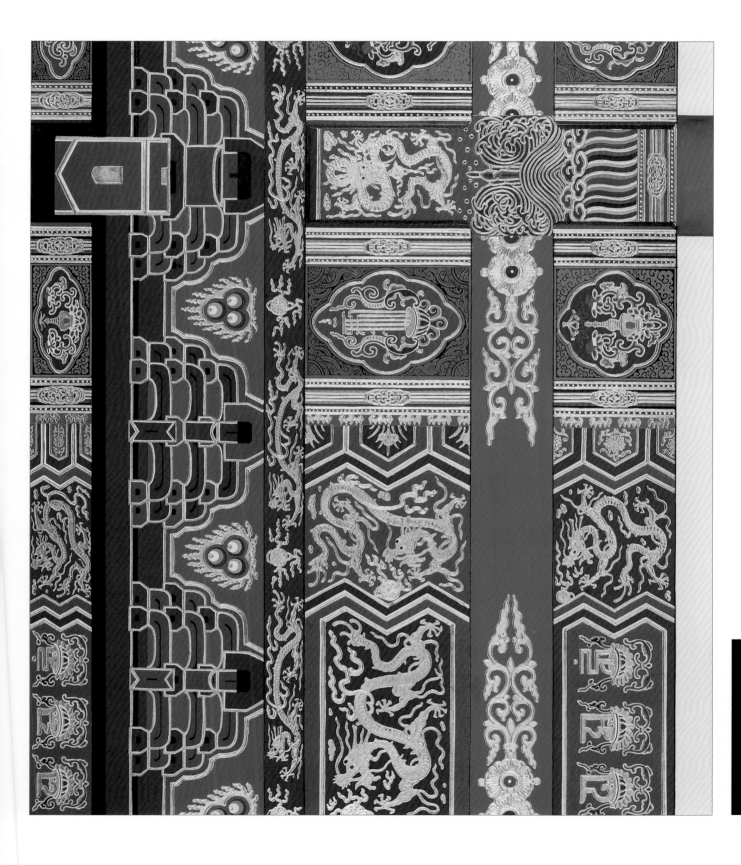

梵文龙和玺彩画　冯世怀

龙草反搭包袱彩画　冯世怀

旋子彩画简述

赵双成

Xuanzi(spiral flower) Pattern

The *xuanzi*(spiral flower) pattern was the principal type of the official architectural decorative polychrome paintings of the Qing Dynasty. The socalled "*xuanzi*" is "*xuanhua*" (spiral flower), a cluster of flowers consisting of many layers of petals, full of mobility. Such a pottern, used as carly as in the Yuan Dynasty (1271-1368), developed gradually from orderless to orderly. The drawing with spiral flowers as its main part came to maturity in the Ming Dynasty (1368-1644) and was further consummated in the Qing Dynasty (1644-1911). As the Qing Dynasty gave prominence to the supremacy of imperial power and established a strict hierarchy, the *xuanzi* pattern was divided into different grades accodingly. The *hexi*(combined dragon seals) pattern growing out of the *xuanzi* pattern became the only decorative polychrome painting which was used exclusively to embody the imperial power, while the formerly dominant *xuanzi* pattern was reduced to a soconadry rank.

There were 8 grades in the *xuanzi* pattern. They were painted mainly in the palaces, and temples of secondary importance, as well as on the gates and corridors in the imperial palace.

旋子彩画是清代官式建筑彩画的一个主要类别。所谓"旋子"即"旋花"，是由多层次旋动感很强的花瓣组成的一种团花纹饰。旋花纹饰早在元代即有运用，它经过了由无序到有序的不断完善。明代以旋花作为主体纹饰的旋子彩画就已经很规范成熟了，到了清代更进一步得到了充实和发展。

清王朝为了突出其皇权地位的至高无上，自上而下确立了严格的等级制度。因此，旋子彩画也随之划分出了高、中、低多层次的等级做法，并在旋子彩画框架基础上演化出了代表皇权专用的"和玺彩画"类别，而一直占主导地位的旋子彩画下降了一个等级，成了低于和玺彩画类别的第二类别彩画。但高等级的旋子彩画由于构图复杂，做法细腻，用金量大等特点，也不失其庄严华贵、富丽辉煌。

旋子彩画基本上可分为八个等级：

1. 浑金旋子彩画；
2. 金琢墨石碾玉旋子彩画；
3. 烟琢墨石碾玉旋子彩画；
4. 金线大点金旋子彩画；
5. 墨线大点金旋子彩画；
6. 墨线小点金旋子彩画；
7. 雅五墨旋子彩画；
8. 雄黄玉旋子彩画。

在以上八个等级中，除浑金、雄黄玉两种等级做法外，其余六种均以青、绿两色为主色调，雄黄玉则以雄黄色为底色。另外，"旋子加苏画"也是清代以后演变出的一个等级品种，但使用不太广泛。

旋子彩画的构图格式主要分箍头、盒子、找头、方心等部位，但找头部位的花纹不论等级高低必须绘以由旋花组合成的纹饰，其余各部位纹饰则可视等级高低而加以变化。其等级层次由高到低可分为以下三种。

1. 方心纹饰。绘龙纹、锦纹、夔龙纹、卷草纹、花卉等纹饰，最低等级则只画一黑杠压心，专业术语称为"一统天下"方心。

2. 箍头纹饰。分为观头箍头、片金箍头及无纹饰的素箍头（死箍头）。

3. 盒子纹饰。其变化基本同方心，可绘龙纹、异兽、卷草纹、花卉及几何图形的栀花纹饰等。

旋子彩画主要施绘于皇家建筑的次要殿宇、门庑及祭祀的坛庙、殿堂等建筑。

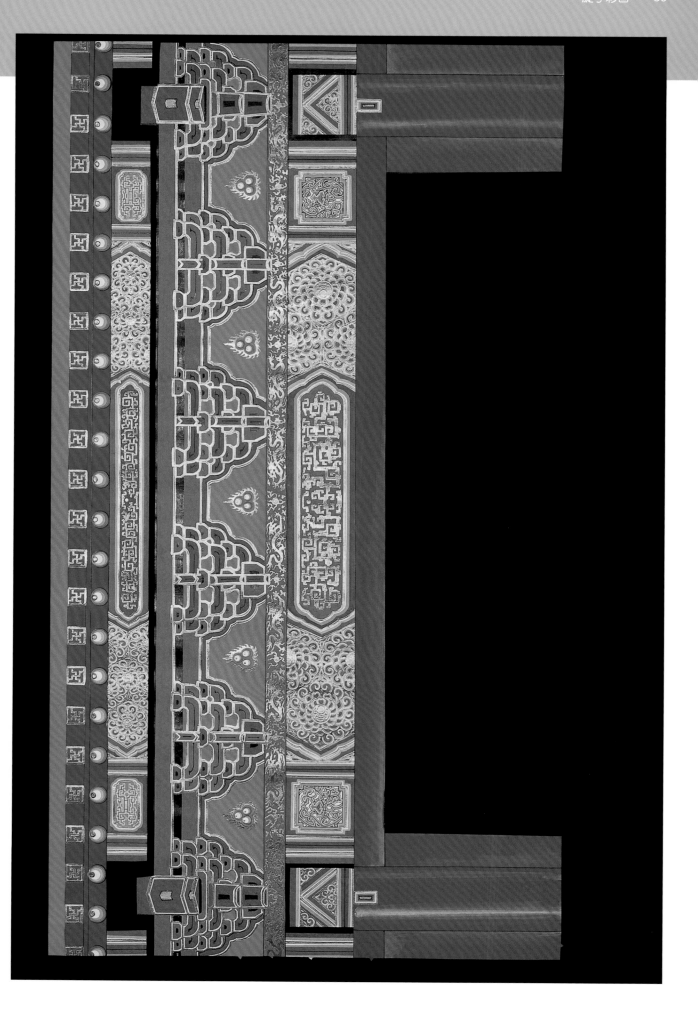

雍龙方心金琢墨石碾玉旋子彩画　蒋广全

龙锦方心金琢墨石碾玉旋子彩画　郑书本

龙锦方心金琢墨石碾玉旋子彩画 边精一

39　旋子彩画

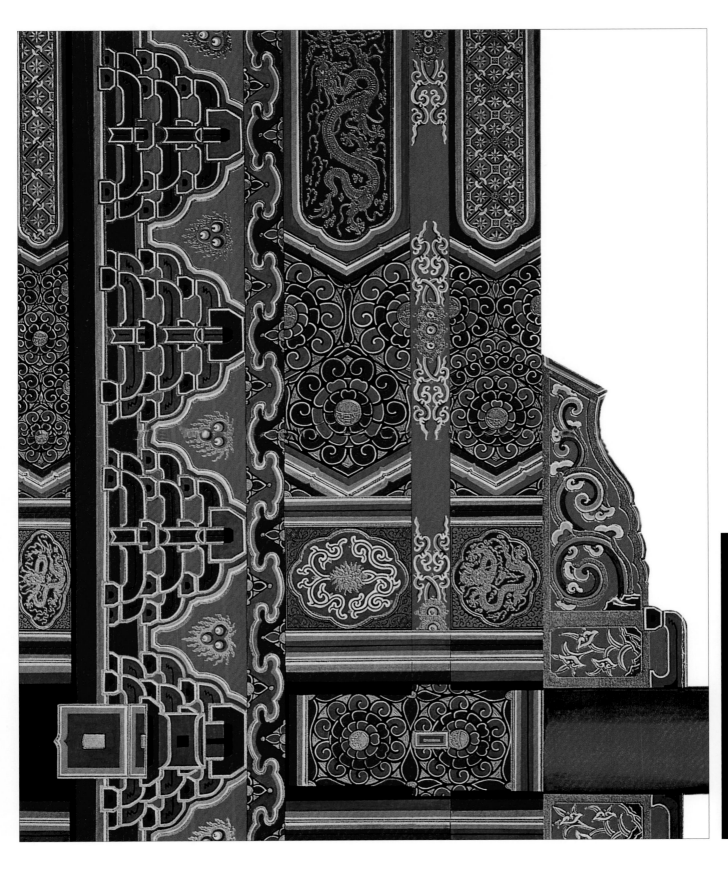

龙锦方心金线大点金旋子彩画　边精一　高成良

金线大点金绘画方心旋子彩画　杜恒昌

41　旋子彩画

龙锦方心墨线大点金旋子彩画　王立新

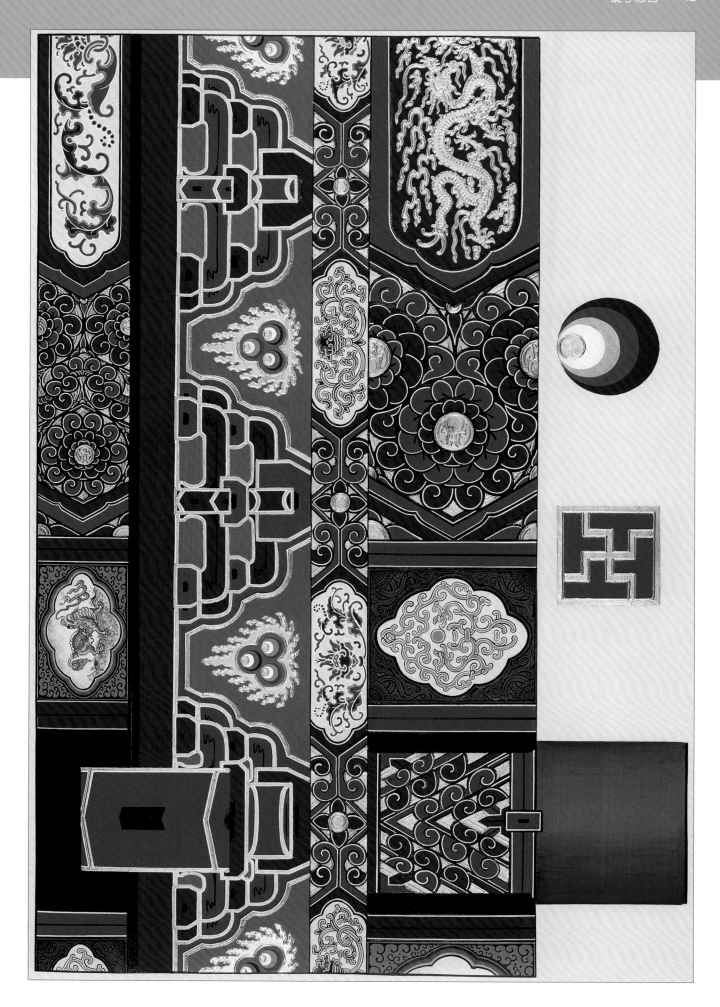

龙草方心墨线大点金旋子彩画 冯世怀

片金西蕃莲宋锦方心墨线大点金旋子彩画 高成良

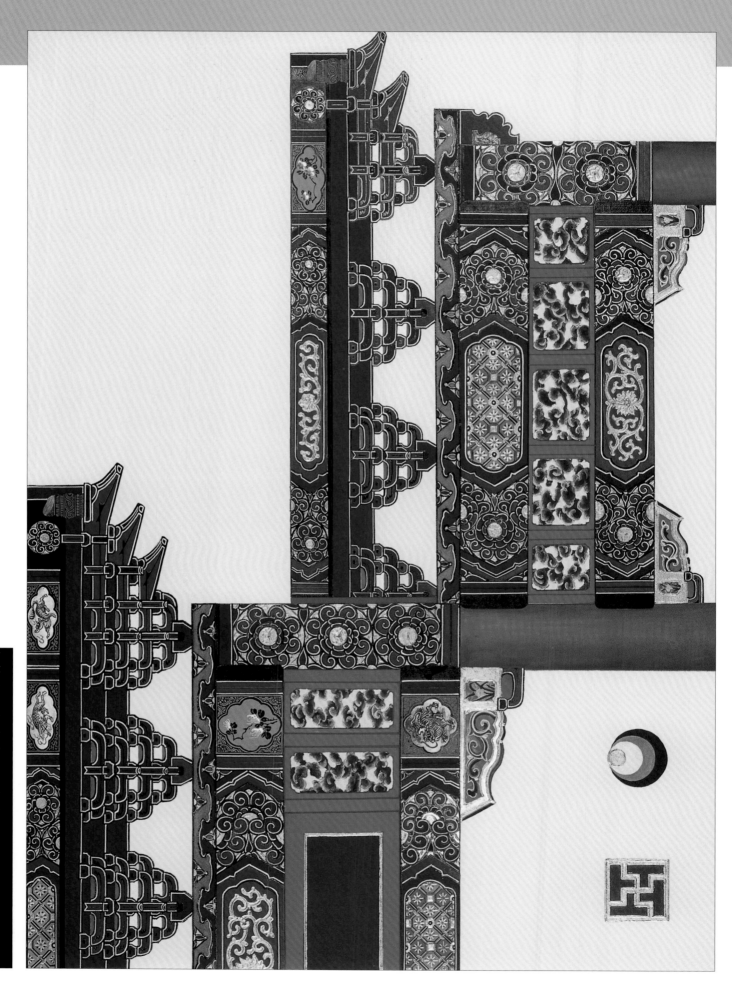

片金西蕃莲末锦方心墨线大点金旋子彩画　冯世怀

襄龙西蕃莲方心墨线小点金旋子彩画　北京文物整理委员会

夔龙西蕃莲方心雅五墨旋子彩画　冯世怀

夔龙西蕃莲方心雅五墨旋子彩画　冯世怀

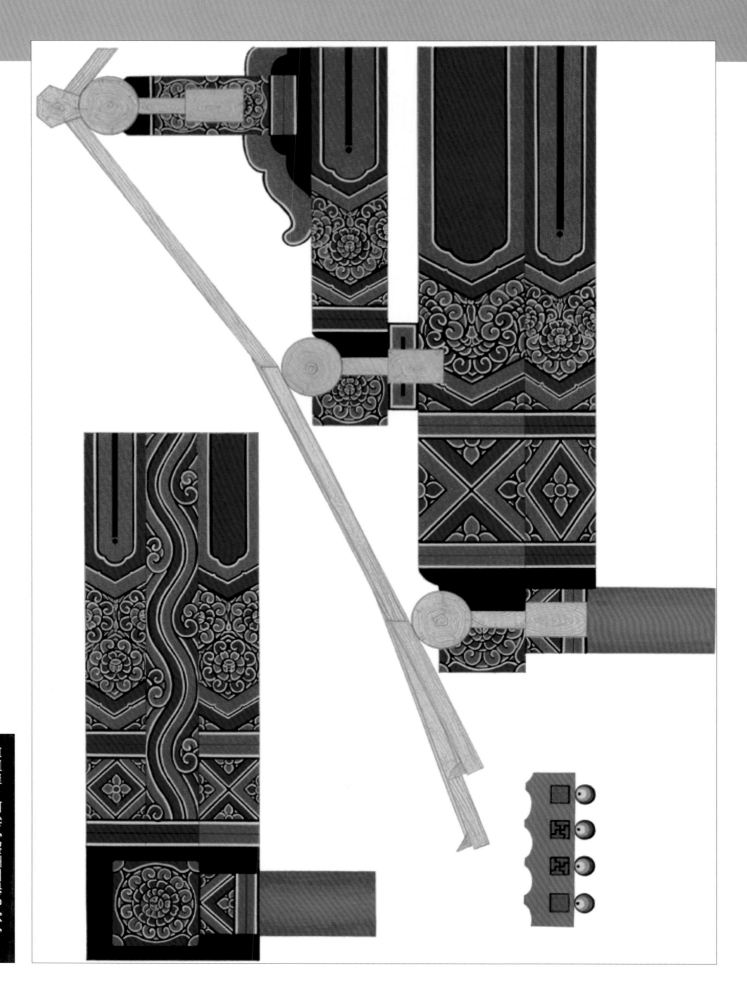

一字方心雅五墨旋子彩画 杜恒昌

49　旋子彩画

夔龙花卉方心雄黄玉旋子彩画　高成良

一字方心雄黄玉旋子彩画　赵双成

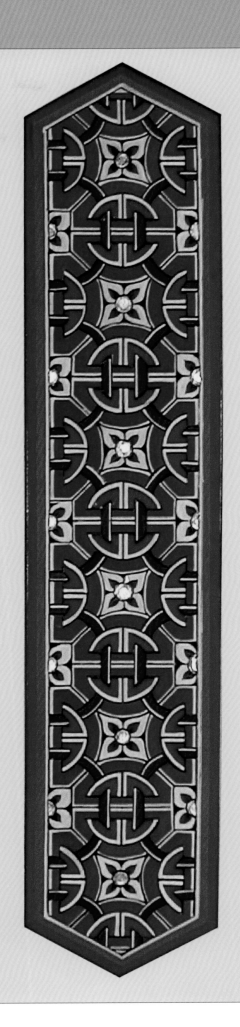

旋子彩画锦纹两则　王立新

旋子彩画锦纹两则 王立新

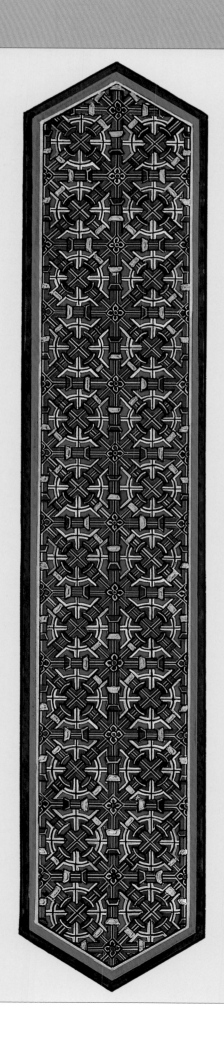

苏式彩画简述

杜恒昌

苏式彩画是装饰园林建筑的一种彩画。它源于江南苏州一带，传至北方进入宫廷即成为官式彩画中的一个重要品种。目前，我们所见到的官式苏画最早是清中期时的彩画，从总体构图至细部纹饰已经完全官式化、北方化，很难看出苏州彩画的痕迹了。苏州地区彩画从纹饰到色彩颇为活泼，以追求素雅无华为基本风格，且很少用金；而北方官式苏画色彩比较艳丽，以青绿色为主色，同时根据装饰内容的需要配以相当数量的间色和用金。

清代官式苏画大体上分为早中期官式苏画和晚期官式苏画。苏式彩画从构图分析大体可分为三种，即方心式、包袱式及海墁式。早中期方心式苏画的主体构图与旋子彩画的主体构图是相似的，找头部分做了变动，换上了锦纹、团花、卡子、聚锦一类图案，方心部分基本绘制龙纹、凤纹、西蕃莲等纹饰。包袱式苏画的找头部分画法与方心式基本一致，包袱内多绘"寿山福海"、"海屋添筹"等一类吉祥图案。海墁式苏画梁枋的两端施箍头，箍头的内侧绘双卡子，卡子之间绘卷草纹，蝠磬纹或黑叶子花，写生画只占很小的比例。

早中期苏式彩画大致可分两个等级。

1. **高等级苏式彩画**。其主体线路为金线，细部纹饰为片金或金琢墨攒退做法。

2. **低等级苏式彩画**。其主体线路为墨线，细部纹饰为烟琢墨攒退局部施金做法。

晚期苏式彩画和早中期苏式彩画相比较，在类别上没有大的变化，依然是三种格式，即包袱式、方心式和海墁式，但细部纹饰方面有了很大变化。比如：方心式烟云叉口和包袱式烟云边框，由早期的花纹边框和烟云边框过渡为同一造型的烟云边框，并增加了烟云托子。箍头由原来的素箍头为主演变成带有各种内容的活箍头，万字、回纹箍头是常用的做法。找头部分主要施卡子、聚锦、黑叶子花卉、博古、池子等，包袱、聚锦、池子内的纹饰演变成以写生画为主的内容，画山水、人物、翎毛花卉、楼台殿阁及传奇故事等。包袱、方心、聚锦底色由原来的重彩色改为白色或浅淡色，称之"白活"。方心式苏画晚期特征与包袱式是一致的。海墁式的主体框架构图无大变化，突出变化是以流云纹和黑叶子花卉为主要内容。

晚期苏画分为高、中、低三个等级，高级做法是"金琢墨苏画"，中级做法是"金线苏画"，低级做法是"黄线苏画"。等级的差别是通过纹饰的用金量大小、工艺绘制的繁简而体现的。

苏式彩画多用于皇家园林或楼、阁、榭、轩、游廊等建筑中。苏画的构图设色富有变化，色调清雅活泼，与幽静的庭院环境相映成趣，别具风格。

清代以后，苏式彩画的图案内容更加丰富，应用范围更广了。

Suzhoustyle Pattern

The Suzhoustyle Pattern is a kind of polychrome paintings used to decorate landscape gardens. It originated in Suzhou area and became an important genre of the official decorative paintings after it came into the imperial court. Up to now the earliest official Suzhoustyle pattern we have ever seen is the one painted in the middle period of the Qing Dynasty. From the general composition to the detail pattern it was official without any trace of the indigenous Suzhoustyle. Both the pattern and color of the Suzhoustyle pattern are lively. Its basic style is plainness and elegance, but the norther official decorative paintings are brigh colored, with blue and green as dominant hues.

The official Suzhoustyle pattern of the Qing Dynasty came in two classes, those in the early and middle periods and those in the late period.

The Suzhou style pattern of the early-middle period have two grades, high and low grade, and those of the late period have three grades, high middle, and low.

The Suzhoustyle pattern is mostly used in gardens, verandas and other minor structures.

After the downfall. of the Qing Dynasty in 1911, the Suzhou style pattern kept developing forward. The pattern is even richer, and the area of application is even wide.

苏式彩画 54

清中期方心式苏画（一） 园林古建公司

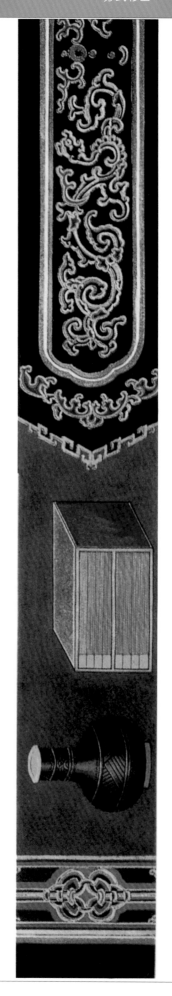

55 苏式彩画

清中期方心式苏画（二） 杜恒昌

清中期方心式苏画（三） 杜恒昌

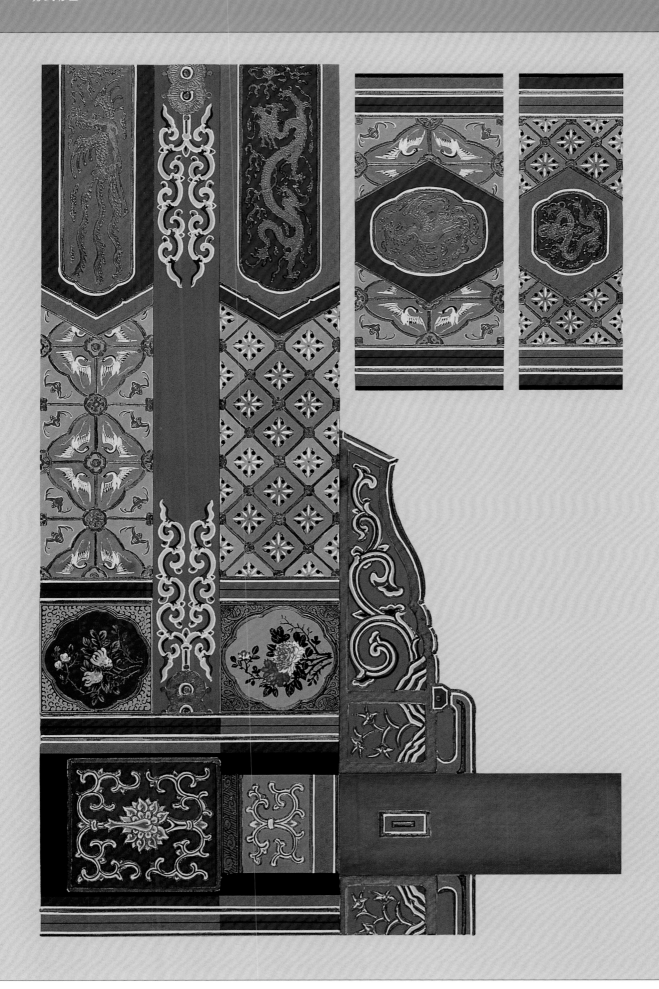

清中期方心式苏式彩画（四）　杜恒昌

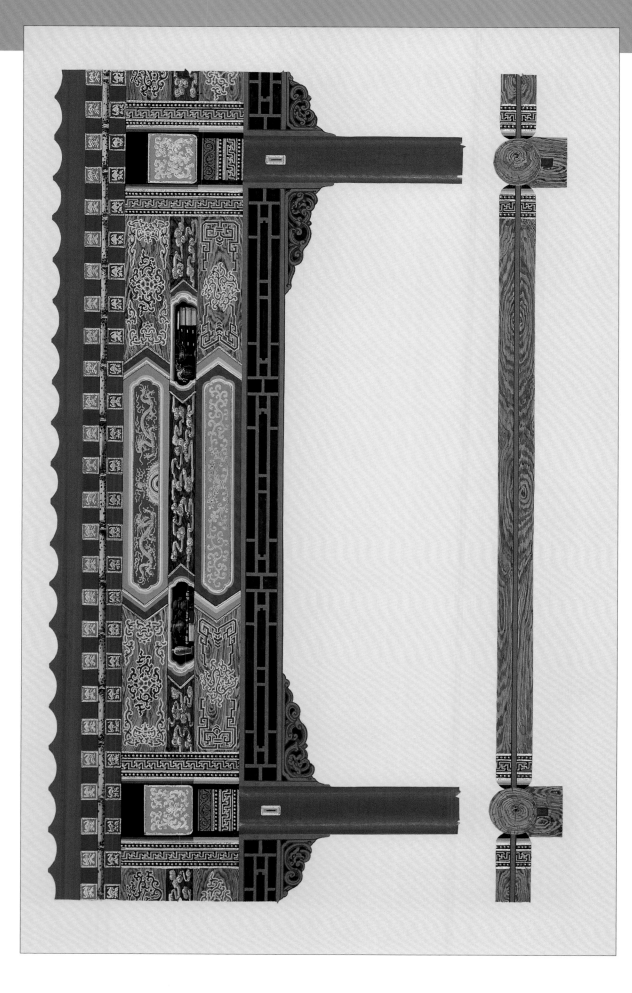

清中期方心式云秋木苏画(五) 蒋广全

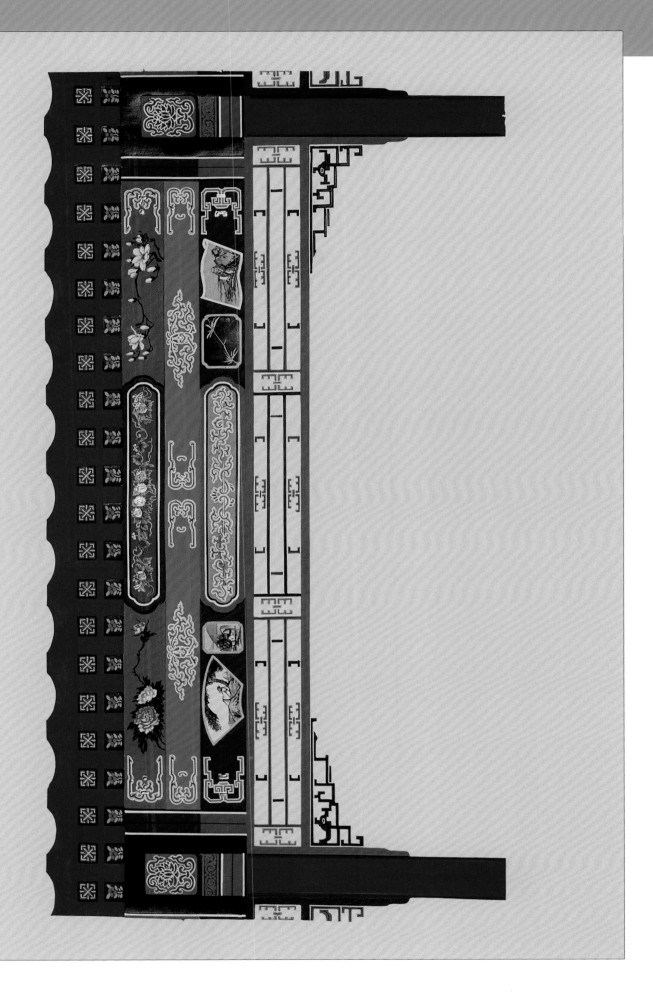

清中期方心式苏画（六） 蒋广全

苏式彩画　60

清中期方心式苏画（七）　园林古建公司

61　苏式彩画

清中期包袱式苏画（八）　园林古建公司

清中期包袱式苏画(九) 蒋广全

清中期包袱式苏画(十) 蒋广全

墨线海墁锦纹双蝠葫芦团花苏式彩画　　蒋广全

65　苏式彩画

清中期金线海墁流云黑叶花卉苏式彩画　蒋广全

清中期海墁式苏画 冯世怀

苏式彩画

清晚期方心式苏画 杜恒昌

清晚期方心式苏画　冯世怀

清晚期方心式苏画 冯世怀

金琢墨方心式苏画 边精一 杜恒昌

71　苏式彩画

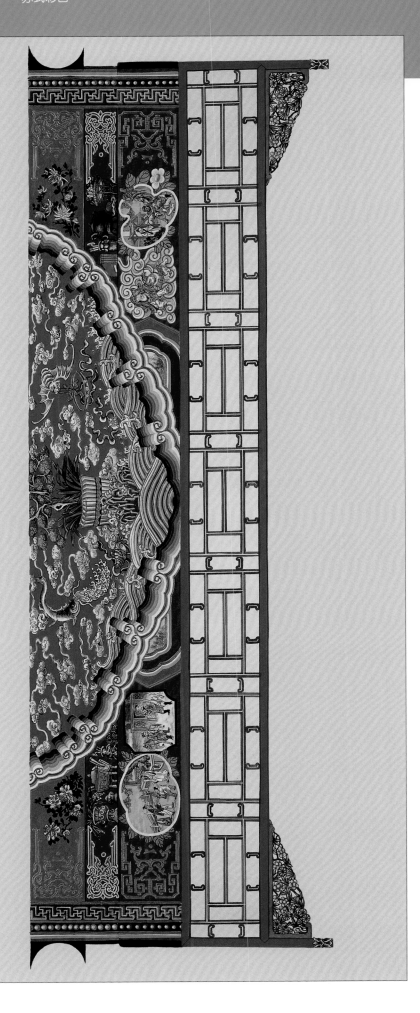

清晚期包袱式苏画　高成良

苏式彩画　72

包袱式苏画　杜恒昌

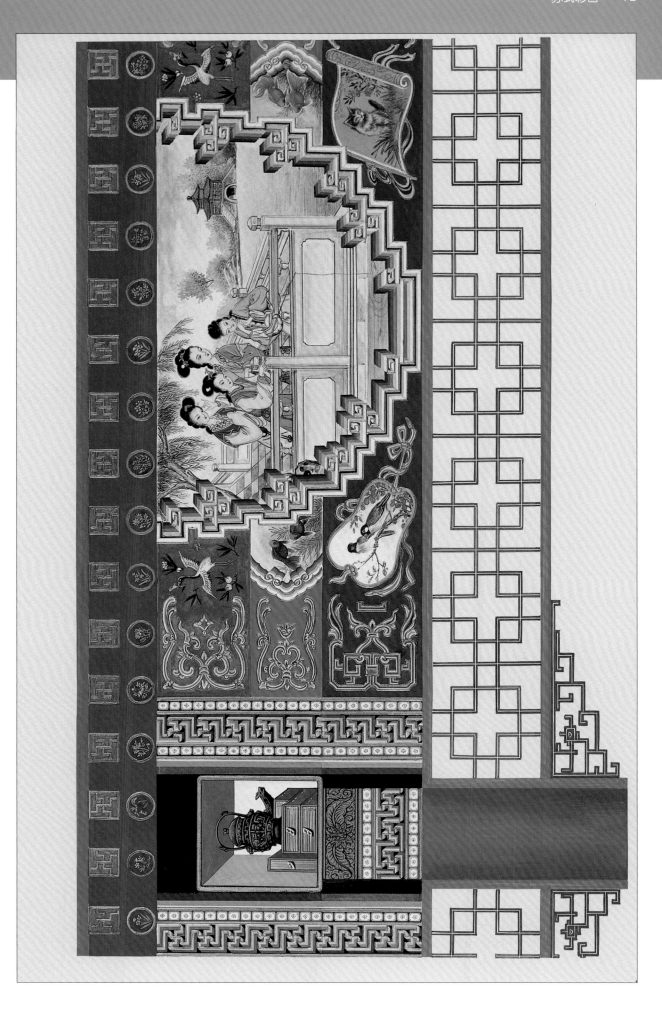

汉瓦箍头包袱式苏画 边精一

清晚期包袱式苏画 蒋广全 赵双成

75　苏式彩画

清晚期包袱式苏画　冯世怀

清晚期挂檐方心式苏画　高成良

苏式彩画 77

掐箍头搭包袱苏画 杜恒昌

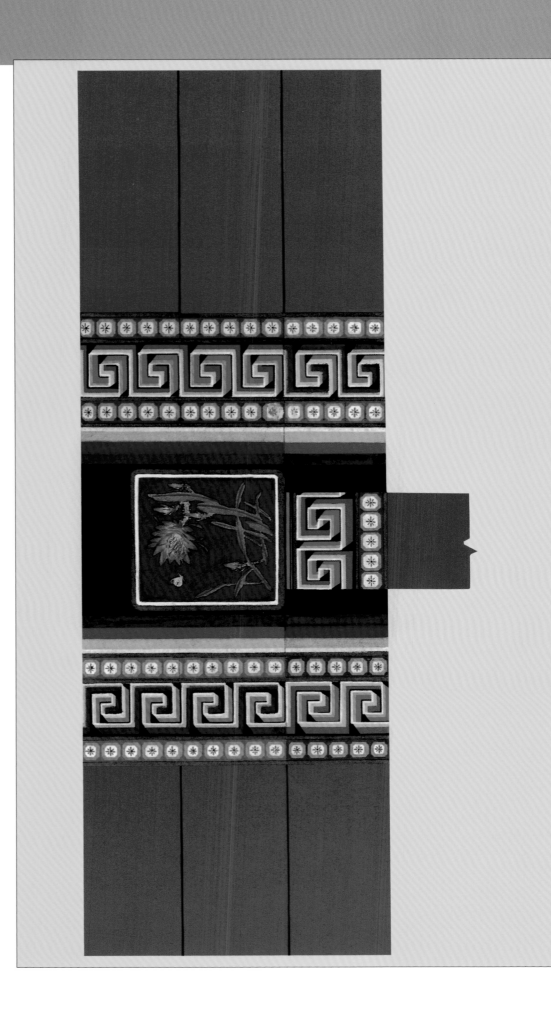

抱框头彩画　赵金城

苏式彩画

拍箍头彩画　赵金城

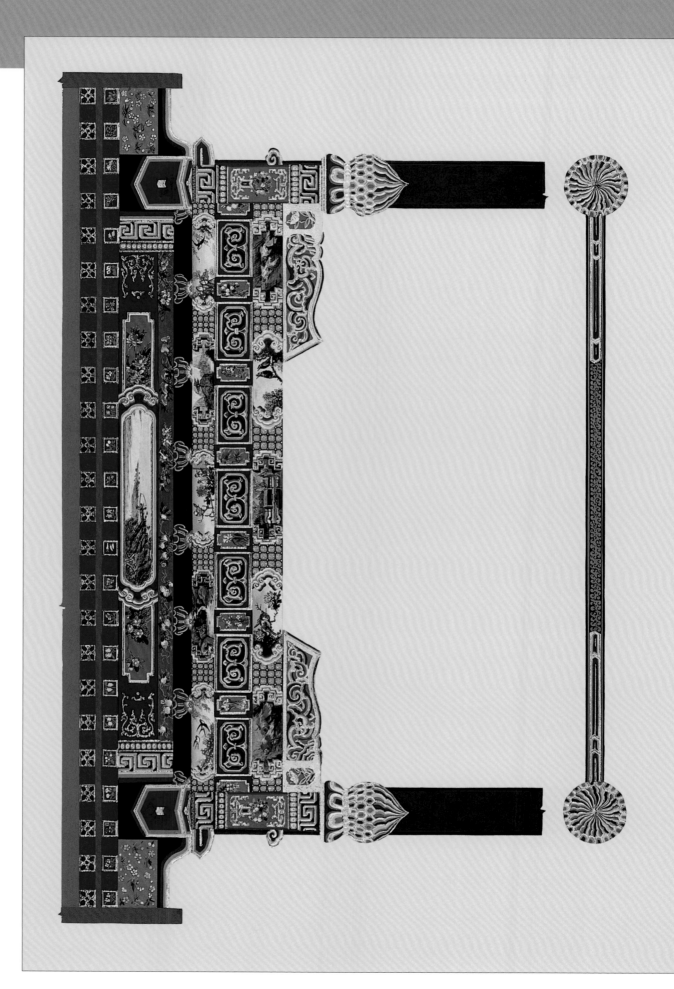

金线垂花门方心苏画　蒋广全

苏式彩画

墨线垂花门方心苏画 蒋广全

苏式彩画　82

包袱聚锦彩画　杜恒昌

苏式彩画

落墨搭色人物包袱彩画　蒋广全

苏式彩画 84

包袱聚锦彩画 高成良

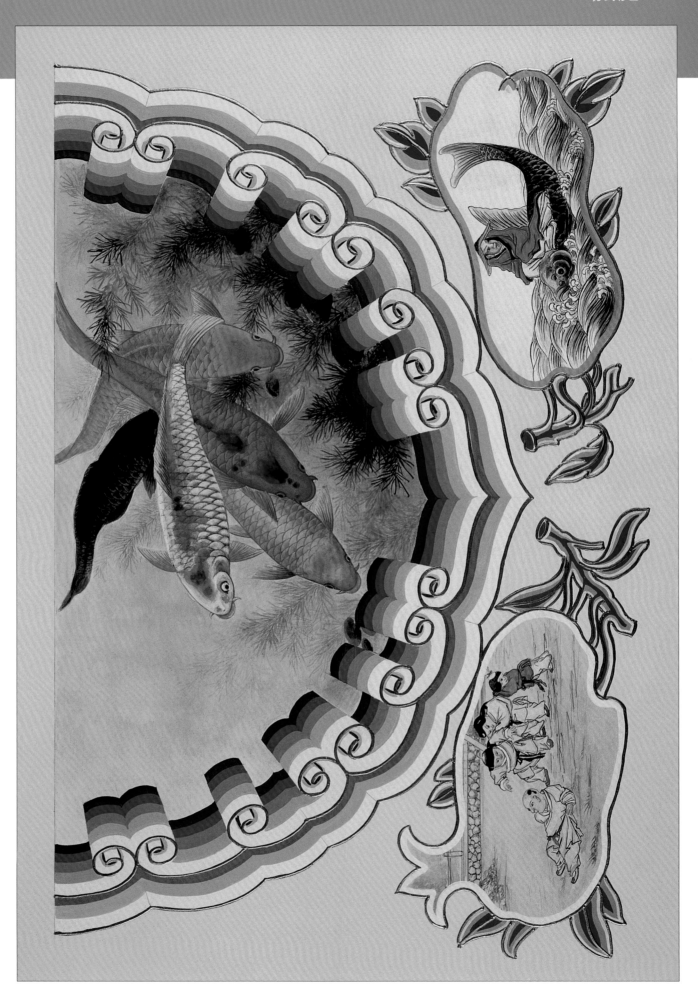

包袱聚锦彩画 杜恒昌

苏式彩画

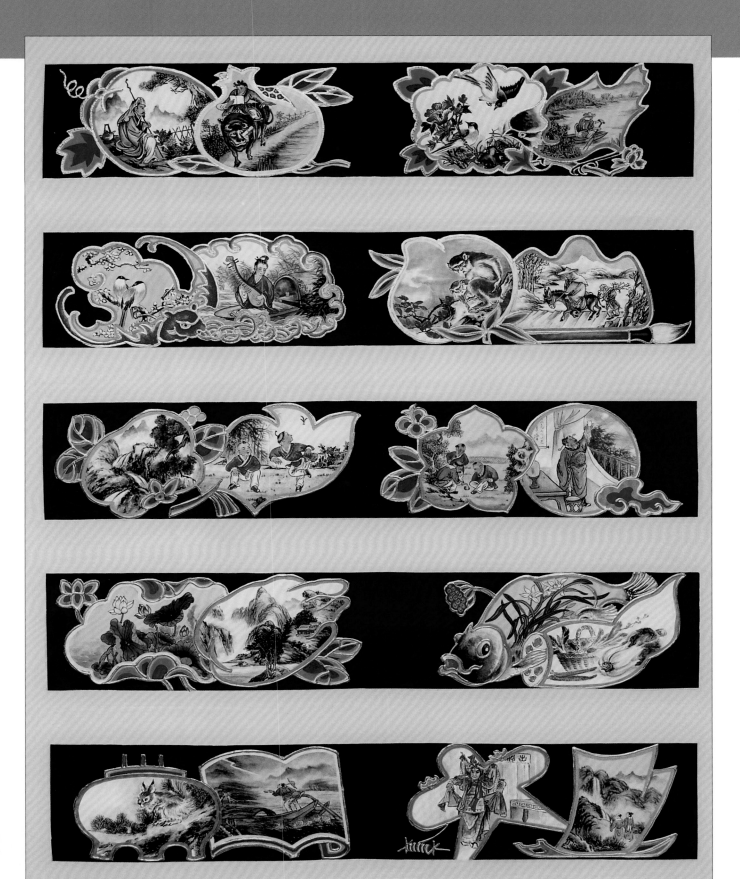

聚锦彩画　高成良

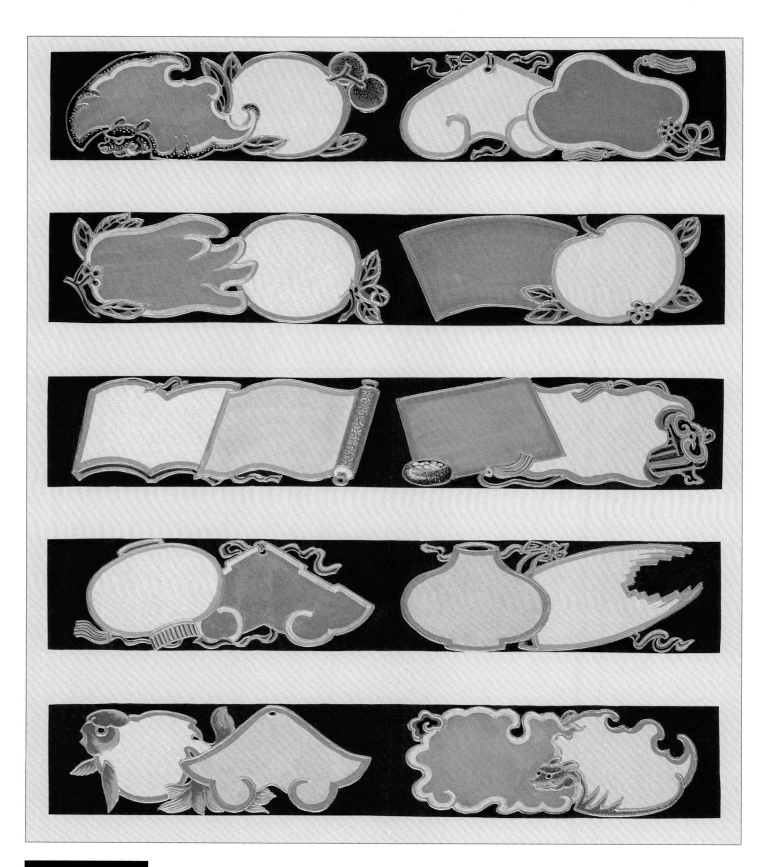

聚锦壳　高成良

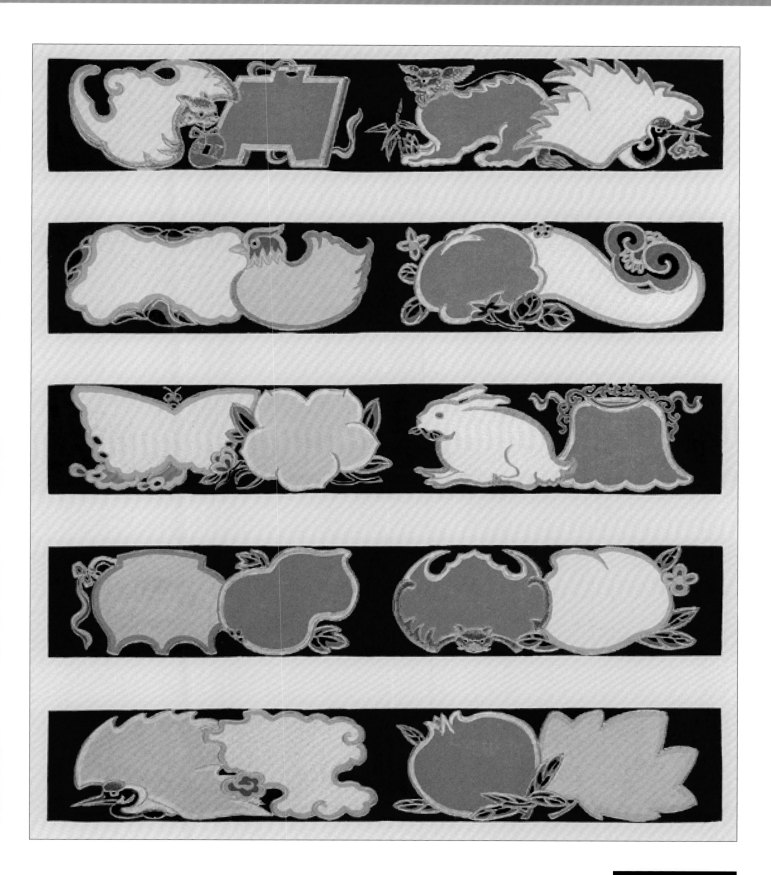

聚锦壳　高成良

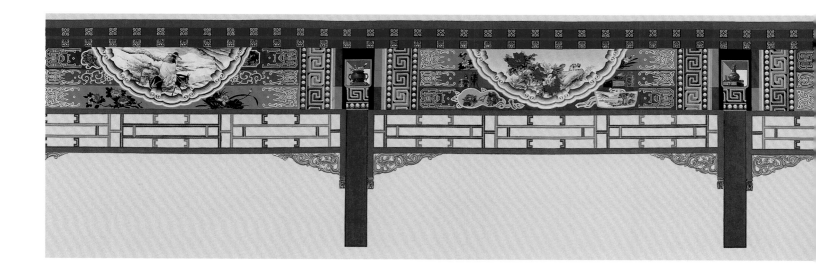

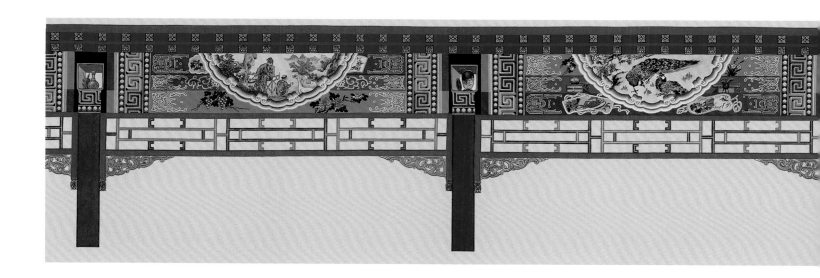

游廊包袱式苏画图卷(1)　北京古代建筑公司集体创作

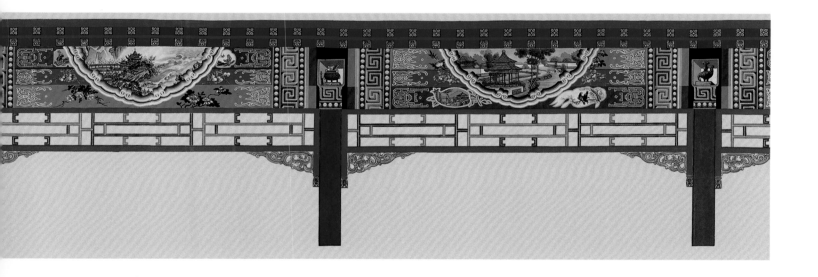

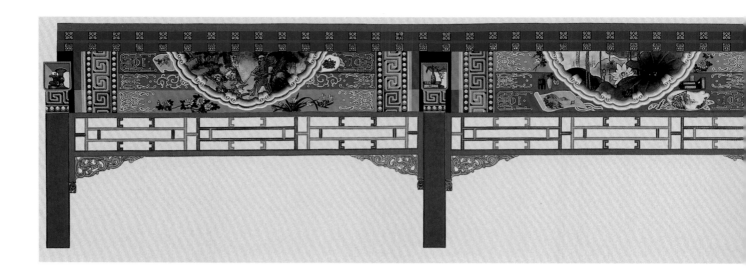

游廊包袱式苏画图卷(2)　北京古代建筑公司集体创作

苏式彩画

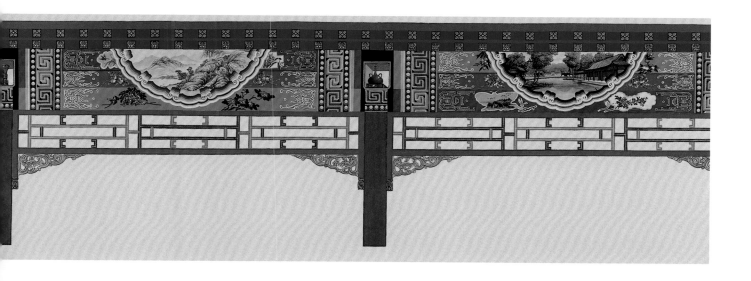

宝珠吉祥草彩画、海墁（斑竹座）彩画简述

Pearl and Lucky Grass Pattern

The pearl and lucky grass pattern is a pattern with a unique style of the variety of the decorative polychrome paintings in the Qing Dynasty. Its composition is so concise that only three pearls as the theme are painted on large wooden components. Around the pearls are huge clusters of flowers consisting of scroll grass patterns. The composition of the pearl and lucky grass pattern is concise, unfolding and with a distinctive style.

The coloring of the pearl and lucky grass pattern is unique. The ground hue on the large wooden component is vermilion. the outer rim of the pearl is painted golden. The scroll grass patterns are blue and green. Its tone is flaming, simple and unsophisticated.

Judgied from the composition and coloring of such a pattern, the pearl and lucky grass pattern, rich in the flavour of the Manchu and Mongolian nationalities may come from Northeast China. It was seldom seen after the middle period of the Qing Dynasty and fused into the official decorative paintings.

Fullcovering (mottled bamboo base) Pattern

Drawings of this kind are different from other types of decorative paintings in the Qing Dynasty. Other type of painfings were painted on components such as large beams and rafters, but the full covering pattern is painted on almost all components from the eaves to the frames of columns.

According to the patterns and color such a pattern is painted in two ways:

1. The mottled bamboo patterns are painted on all components. The bamboos are painted as old and young bamboos and spots are painted on the nodes so as to show that bamboos are mottled ones, hence the name of the mottled bamboo base.

2. The whole building is decorated with polychrome paintings. All components are covered with dark green or light blue oil coatings on which are painted climbers (tendrils) winding along the components. On the lower part of the columns in some buildings the painting is in the form of rocks from Tai Lake. Such decorations are to show the beautiful natural scenery.

Judging from the existing decorative paintings such paintings might originate in the late period of the Qing Dynasty. They were used in limited areas. For example, they were painted on part of the imperial buildings in the imperial gardens and gardens of princes and high-ranking officials.

一、宝珠吉祥草彩画

宝珠吉祥草彩画是清代诸多彩画中一个非常独具风格的品种。这类彩画的构图极其简练，梁枋上面不设方心、盒子等基础框架格式，只在大木构件的中心部位绘出三颗宝珠纹作为主题。宝珠周围用卷草纹组成硕大花团形纹饰，构件两端设箍头，箍头的内侧上端用卷草纹组成近似岔角形的纹饰。宝珠吉祥草彩画的构图简洁、舒朗，别具一格。

宝珠吉祥草彩画的设色也极有特色，大木构件是以朱红色作为底色，宝珠的外缘用金色装饰，卷草纹用青绿色，并点缀金色，这类彩画的色调极其炽烈古朴。

从这类彩画的构图和色调分析，极具满蒙民族建筑装饰的风范，很可能是从关外流传至京城的。清代中期以后此种彩画已不多见，渐渐地与官式彩画融于一体，以新的形式出现于建筑之上。

二、海墁（斑竹座）彩画

这类彩画不同于清代的其他类别彩画。一般的彩画所装饰部位主要集中于大木梁枋、斗拱、椽飞等构件，下架柱框装修通常采用油饰处理。海墁（斑竹座）彩画施绘彩画的范围扩大到一座建筑的几乎所有木构件，上至连檐瓦口，下至柱框。

从纹饰和色彩方面划分，这类彩画大致有两种做法。

其一是所施绘彩画的构件遍绘班竹纹。构件表面绘出一排排组合有序的竹竿纹，宛如用纤细的竹竿搭建而成。竹竿绘成"老竹"和"嫩竹"两种质感，竹节处点染斑纹，以此显其竹皆为斑竹（匠师们也称此种做法为斑竹座）。

其二是整座建筑施绘同样彩画。凡构件遍涂深绿或淡青色的油皮，其上绘出缠绕构件生长的藤萝等花卉，有的建筑在柱的下部绘出太湖石造型，此种装饰的意境是描述自然景观之美。

从所存的实例分析，这类彩画可能产生于清代晚期，使用范围也很有限，多见于皇家园囿和王公大臣花园中的部分建筑。

95　宝珠吉祥草彩画　海墁（斑竹座）彩画

清代宝珠吉祥草彩画　海墁

宝珠吉祥草彩画 海墁（斑竹座）彩画 96

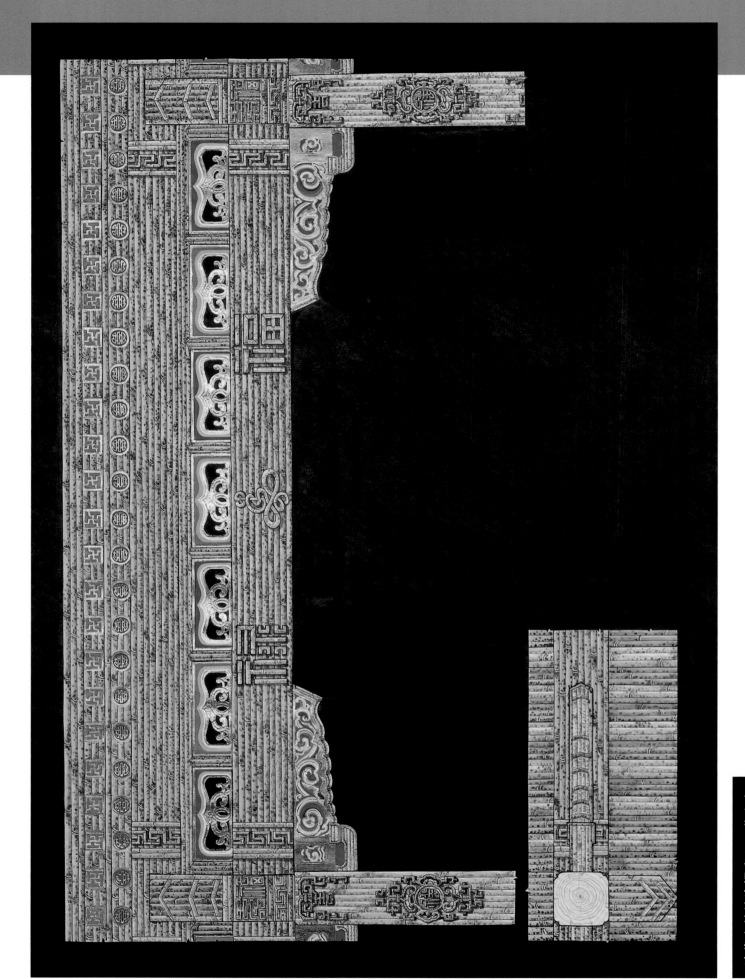

清代斑竹纹海墁彩画 蒋广全

清代官式彩画檩枋大木以外从属构件简述

Brief Introduction to the Qing Official Decorative Polychrome Paintings on the Secondary Components (other than the Wooden Structural Beam)

by Jiang Guangquan

Mainly employed on the wooden structures such as beams, purlins and ties, the various types of the Qing official decorative polychrome paintings are clearly oriented to different hierarchical levels. However, the paintings are also employed on other relating secondary components: rafter ends, flying rafter ends, rafter and roof boarding, bracket sets (dougong), cushion boards, ceilings, stalks, sparrow braces (queti), and wainscots, etc. Since the hierarchy or structure varies in each individual building, not all the above components are indispensable except the rafter ends. In fact, the configuration of the components lies on the setting of the specific hierarchical levels. The building in the highest level, of course, contains every of the secondary components. Here is the general rule for embellishing them: the decoration and relevant manners of the secondary components should match those of the wooden structural beams. Due to the nature of these sorts of decorative polychrome paintings —— the variety in particular decorations yet the conformity in general rules —— here will discuss them in an integral way, without classifying them into different types or hierarchical levels.

(张宇 翻译 / translated by ZHang Yu)

檩枋梁之上所绘彩画是各类彩画的主要标志所在，其类别分明、等级清晰。与之相关联匹配的另一部分构件有：椽头、飞头、椽子望板、斗拱、垫拱板、天花、支条、雀替、墙边等等。由于建筑的等级、结构的不同，不是每座建筑都件件具备，除了椽头是必不可少的以外，其余构件皆因等级而定（最高等级的建筑当然包括以上所列的各种构件）。这些构件的彩画纹饰、设色和做法，总的要求是：与该座建筑大木彩画的等级相匹配，与其纹饰内容、做法层次相呼应。鉴于这部分彩画的纹饰繁多，又具有一定互通共用的特性，不宜逐类一一纳入各类彩画之中，只能单独列出。

清代官式彩画檩枋大木以外从属构件　98

椽头、飞头、椽子、望板彩画　蒋广全

清代官式彩画檩枋大木以外从属构件

斗拱彩画　边精一

清代官式彩画檩枋大木以外从属构件

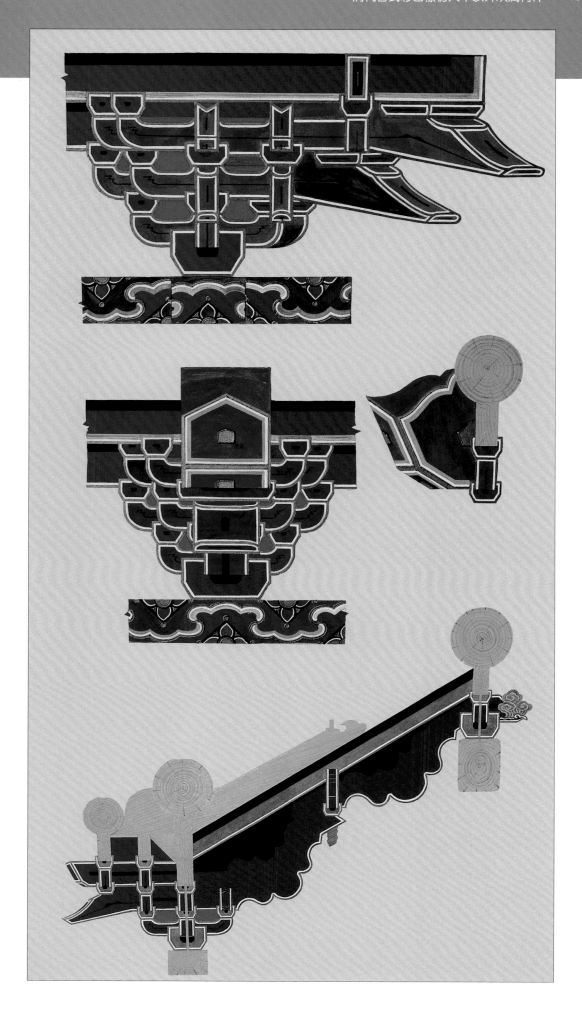

斗拱彩画　边精一

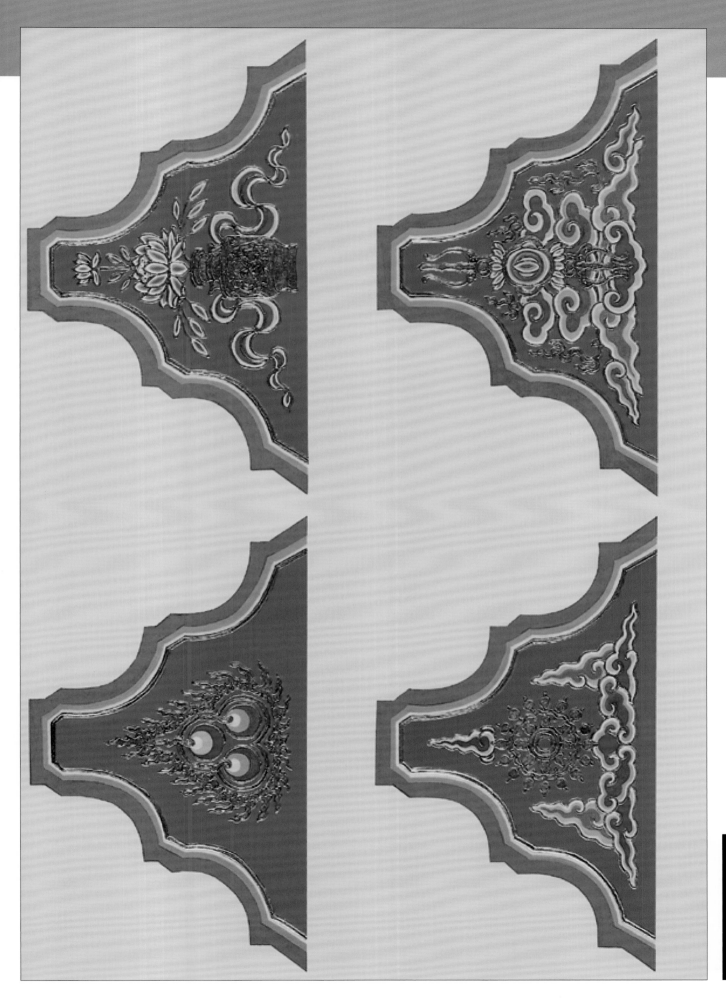

垫拱板彩画 杜恒昌

清代官式彩画檩枋大木以外从属构件　102

片金坐龙天花　杜恒昌

片金坐龙天花　蒋广全

清代官式彩画檩枋大木以外从属构件　104

片金升降龙天花　蒋广全

片金双凤天花　蒋广全

清代官式彩画檩枋大木以外从属构件　106

团鹤天花　蒋广全

双鹤天花　蒋广全

清代官式彩画檩枋大木以外从属构件　108

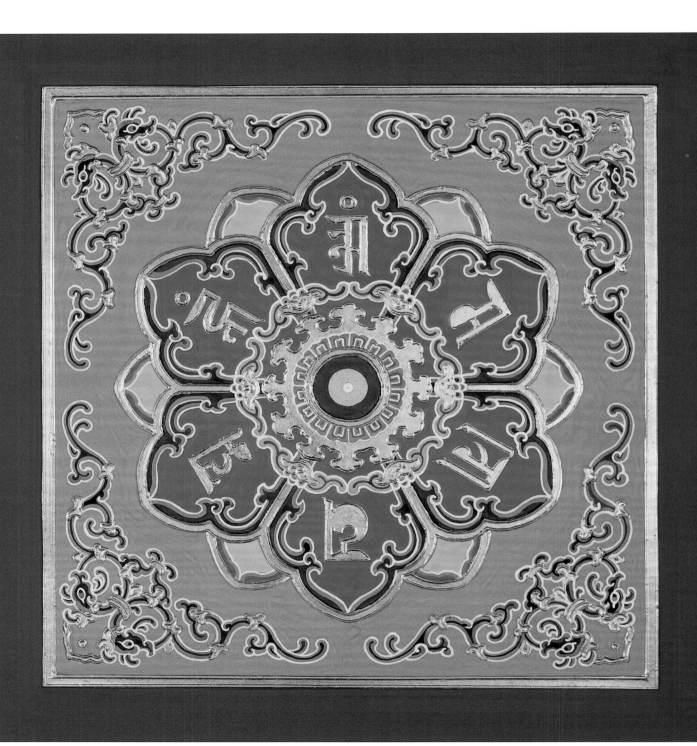

夔龙岔角六字真言天花　蒋广全

109　清代官式彩画檩枋大木以外从属构件

六字真言天花　冯世怀

清代官式彩画檩枋大木以外从属构件　110

坐夔龙天花　蒋广全

攒退硬夔龙天花　蒋广全

清代官式彩画檩枋大木以外从属构件　112

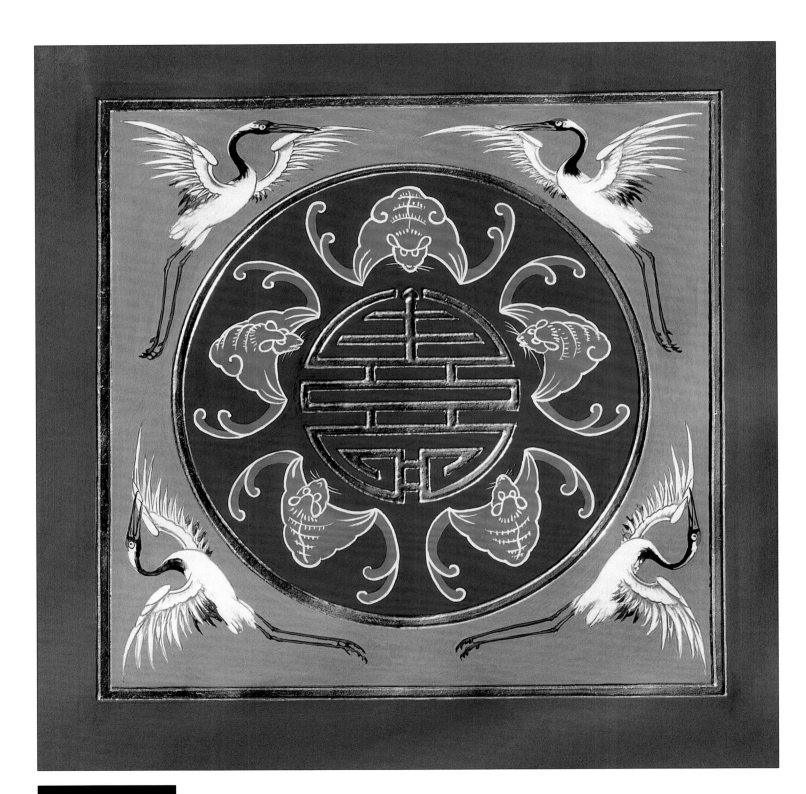

福寿天花　蒋广全

清代官式彩画檩枋大木以外从属构件

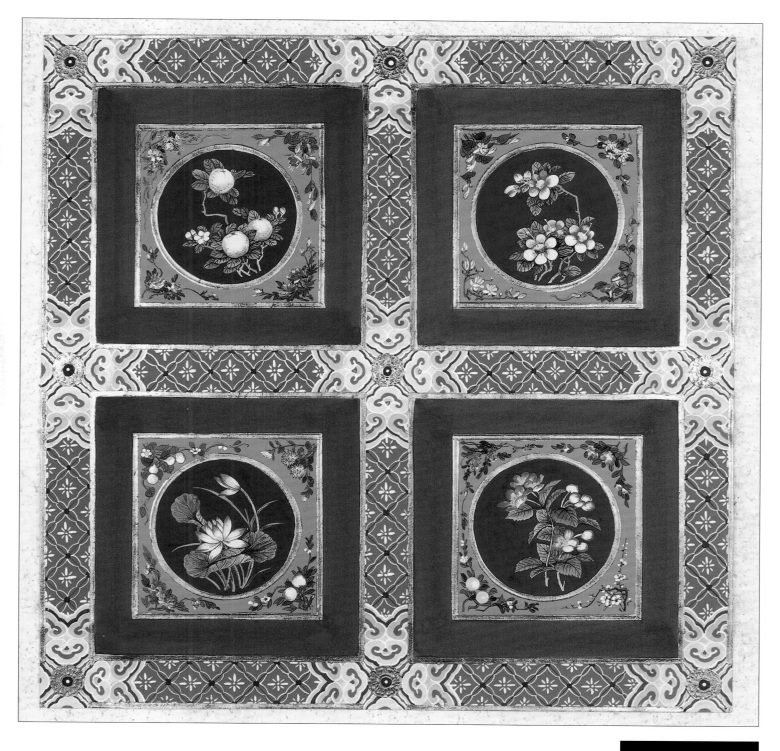

百花天花　蒋广全

清代官式彩画檩枋大木以外从属构件　114

彩龙天花　高业京供稿

115　清代官式彩画檩枋大木以外从属构件

片金西蕃莲天花　蒋广全

清代官式彩画檩枋大木以外从属构件 116

雀替彩画

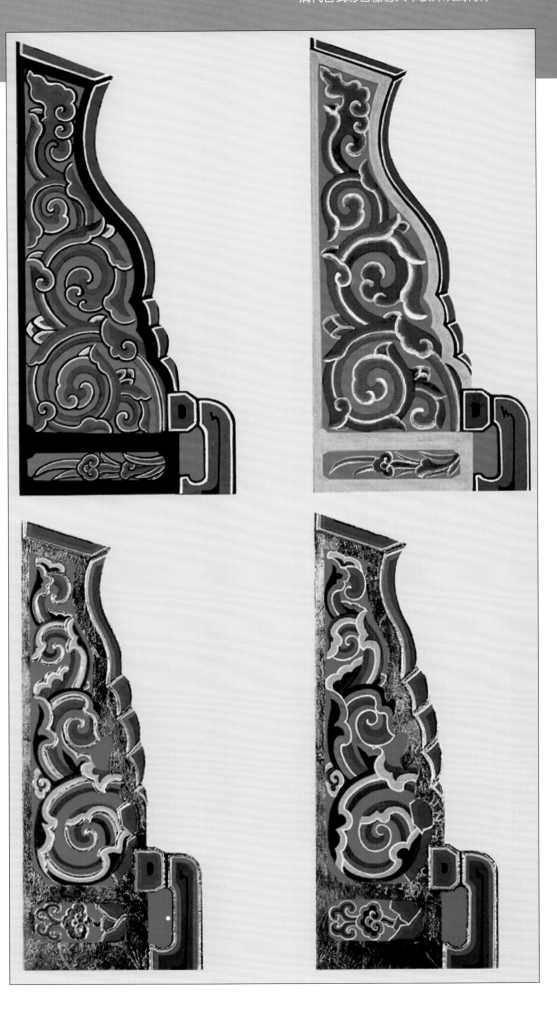

雀替彩画 掭广金

地方彩画简述

Local Decorative Polychrome Paintings

The local decorative polychrome painting is a kind of decorative painting which co-exists with the official decorative painting. They are cognate but different from each other. The official decorative painting with strict hierarchy and standard procedure was painted on imperial and related buildings in different historical periods, while the local decorative painting more lively in form, did not rigidly adhere to standards on the premise of not violating the hierarchy in different historical periods. Both of them have a long history.

Numerous are kinds of the local decorative painting. If classified according to regions there are regions inhabited with Han nationality and regions inhabited with minorities: if classified according to the kind of buildings to be decorated, there are official, religious and folk dwellings. The local decorative polychrome paintings is an essential integral part of the Chinese architectural decorative polychrome paintings. It is the human cultural heritage, too.

The local and official decorative paintings inherited each other in many ways. For example, the patterns of the xuanzi (spiral flower) pattern are similar to those of the local decorative paintings Fig. 1 and Fig. 2 show the patterns of the local decorative paintings popular in the northwestern areas in the Qing Dynasty, and Fig. 3 and Fig. 4 show the official patterns.

As the local decorative paintings is not much fettered by standards and rules, it is lively and full of local characters. The decorative polychrome paintings made in Suzhou and Hangzhou belong to different schools which vary in style. For example, decorative paintings of Suzhou School are different from those made by Hangzhou School. Ningbo School and Huzhou School.

From a macroscopic point of view the difference in decorative polychrome paintings between those made in North and South China is sharper than those made in East and West China.

地方彩画是指与官式彩画并存，既有区别、又一脉相承的一类彩画。官式彩画则是在不同历史阶段适用于当时御用及其相关建筑上面的、做法规范的一类彩画，等级比较严明。地方彩画在不违背当时等级制度的前提下，形式比较活泼，做法不拘泥于程式，它的历史与官式彩画同样悠长久远。

地方彩画种类较多，从区域划分，有汉族地区和少数民族地区之分；从所装饰的建筑类别划分，有衙署建筑、宗教建筑和民居建筑等之分；从所装饰建筑的构造方面划分，有大式建筑和小式建筑之分。地方彩画是我国建筑彩画的重要组成部分，同样是人类宝贵的文化遗产。

由于工作的局限性，我们对地方彩画的了解实在太浅，目前只能粗略地作些介绍。

地方彩画与官式彩画在诸多方面存在着承袭关系。例如清代西北地区旋子彩画的方心纹饰造型（图1）与"喜相逢"旋花造型（图2）；清代官式旋子彩画中的同部位的纹饰造型图3与图4。通过对比不难看出两者有许多近似之处。又如少数民族地区藏传佛教庙宇彩画中的宝珠吉祥草纹饰的造型和绘制方法，与官式彩画中同类纹饰也是极为相似的。这方面的例证很多，本文不再一一列举。

地方彩画由于没有过多的法式规矩所束缚，因而比较活泼自然，地方特色也十分突出。例如苏杭一带的彩画就存在着若干风格各异的派别，计有苏州帮、杭州帮、宁波帮、湖州帮等等。这些帮有的善绘吉祥图案和锦纹，有的善绘戏文，它们的艺术风格迥然不同。

由于所处的地理位置和文化传统的不同，地方彩画的地域特征十分鲜明。从宏观上分析，南北方之间的差别较东西部之间的差别要突出些。如南方彩画侧重于用图案画和编制锦纹，线条一般比较纤细，颜色比较淡雅；而北方彩画喜绘花鸟、人物、山水、器物等写生画，线条一般比较粗壮，颜色崇尚艳丽，浓淡对比明快。

地方彩画是一门亟待我们去挖掘、整理、研究的古代建筑装饰艺术。为了推动此项工作，本集将所积存的几幅样稿编入，实是挂一漏万，望同业人员谅解。

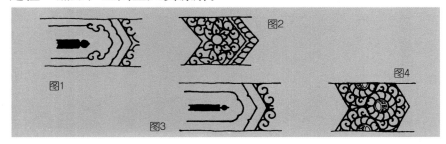

地方彩画 118

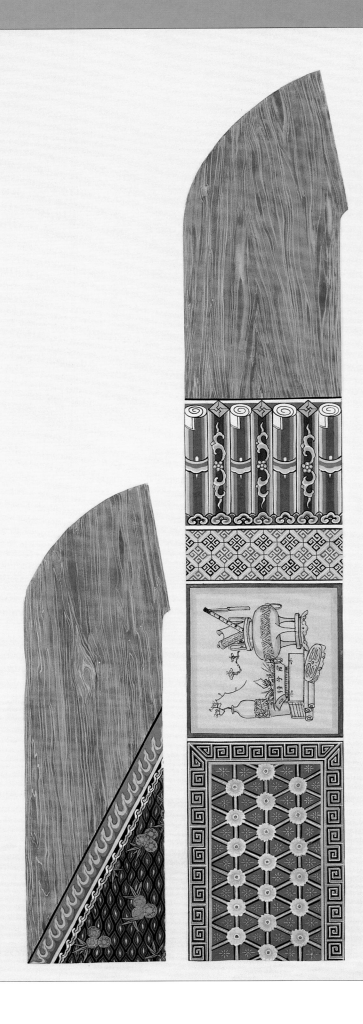

清晚期苏州地方彩画 高成良

119　地方彩画

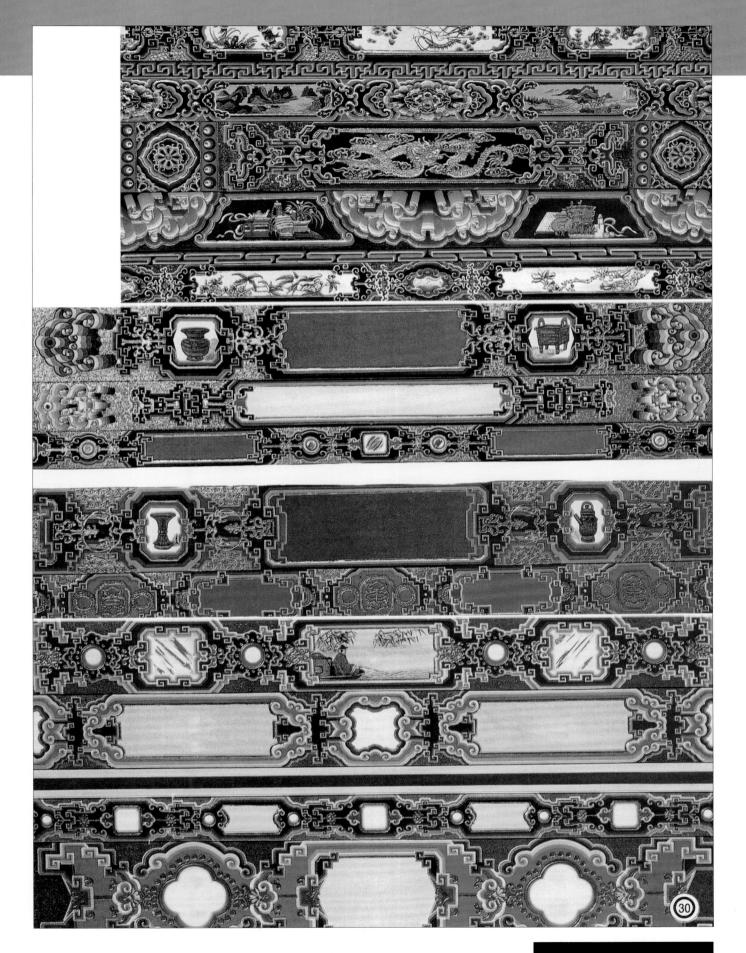

㉚

山西地方彩画　岳俊德等

山西地方彩画　岳俊德等

121 地方彩画

辽宁地方彩画　张世满　李海申

黑龙江地方彩画 郑连祖

123　地方彩画

黑龙江地方彩画　郑连祖

高 成 良

新式彩画简述

辛亥革命以后，北京等大城市中西样式结合的建筑相继出现。如建造于本世纪初的协和医院、北京图书馆等，这些建筑物的出现必然带来建筑装饰彩画的变革。

最初的新式彩画只能说是在摸索中尝试的，格局定式上变化较小，图案套用宋、明、清式传统彩画及纹饰为蓝本，变化不大。使用的色彩在内外檐上稍有区分，优其是室内部分，在色彩运用上不太注重环境和使用功能，色调较为沉闷。外檐则多以传统彩画的纹饰另行组合而已。这是新式彩画刚一出现时的特点。

建国以后，随着新式建筑的不断涌现，给设计人员和彩画匠师们提供了广阔的创作天地，作品逐渐增多。可以说，新式彩画是现代建筑的装饰需要。在继承传统的油饰彩画工艺基础上，同时结合现代工艺和表现手段，吸收各民族建筑装饰艺术之精华，更注重建筑的环境及使用功能的统一，因而使新式彩画日趋完美。

新式彩画没有严格的等级规定，因为它不受"法式"的约束。根据不同环境、部位、功能的需要设计施绘，表现手法灵活多样。一般习惯上以用金量多少、线条与设色层次繁简来区分级别。

新式彩画大多装饰于天花板、灯花、柱子、梁枋、墙边等部位。它是一种以花纹、图案进行造型组合，时代感很强的建筑装饰彩画。

New-type Decorative Polychrome Paintings

Following the Revolution of 1911, appearing in Beijing and other big cities were new buildings in the form of combining Chincse architecture with Western architeeture, such as Xiehe Hospital, Beijing Library, and others. The appearance of these buildings ushered inevitably changes in the decorative drawings.

The early new-type drawings were tentative in nature. There were minor alterations in pattern and the designs were based on the traditional specimens of the Song, Ming and Qing Dynasties. A slight differencc was in the use of colors, especially the tone in the indoor part was rather dull owing to little attention paid to the function and circumstances. This is the features of the time when the new-type Decorative Polychrome Paintings made its first appearance.

After the founding of the People's Republic of China the increasing constructinon of new buildings provides designers and painters with abroad seope of creation, and the number of new works is increasing. The new decorative paintings are the necessity in the decoration of modern buildings. On the basis of inheriting the traditional technology of the colored decorative paintings and combining modern teehnology and means of expression the new decorative polychrome paintings is improving day by day by absorbing the cream of the decorative arts of different nationalities and placing stress on the unity of the environment and functions of use.

There are no strict hierarchical rules for the new decorativ e paintings because they are not restricted by the "Standards", but designed and painted as required by different circumstances, positions and functions of the buildings.

The new-type Decorative Polychrome Paintings is used on the ceilings, columns, rafters and wall edges, It is a decorative drawing comprised of figures and patterns with strong sense of times.

新式彩画——藻井　郭汉图

新式彩画——墙边 郭汉图

新式彩画——墙边　郭汉图

新式彩画——墙边　郭汉图

129 新式彩画

新式彩画——双龙、双凤天花 郭汉图

新式彩画——灯花　赵双成

新式彩画——灯花　杜恒昌

新式彩画——灯花　赵双成

新式彩画——灯花　蒋广全

新式彩画——灯花　蒋广全

新式彩画——灯花　罗翰秋

新式彩画 136

新式彩画——灯花 杜恒昌

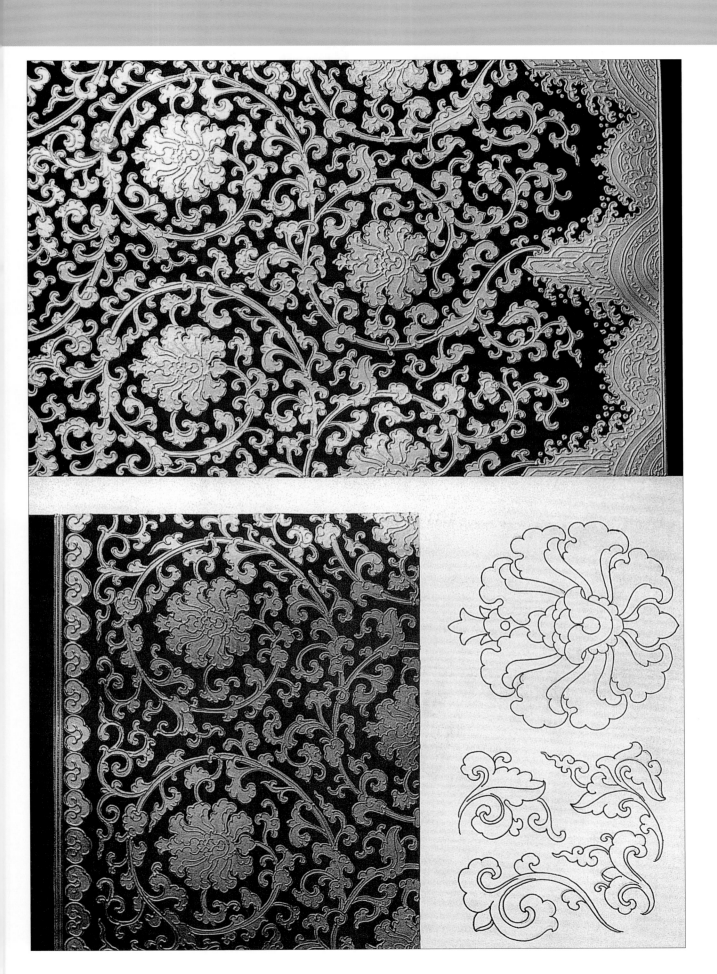

新式彩画——柱 郭汉图

新式彩画——柱　赵双成

新式大木彩画 草汉图

新式彩画 142

新式大木彩画　杜恒昌

143　新式彩画

壁画——八仙醉酒　冯世坂

新式彩画 144

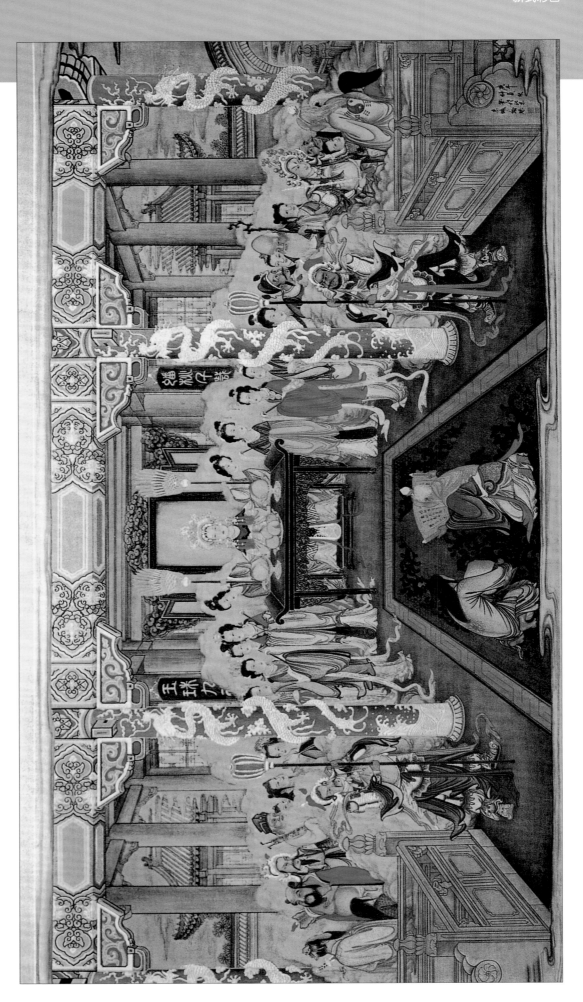

壁画——瑶池盛会　冯世怀

145　新式彩画

壁画——太白醉酒　冯世怀

历代帝王庙景德崇圣殿明间内檐脊部彩画　蒋广全　摄影

故宫景仁宫龙凤方心西蕃莲找头和玺彩画　蒋广全　摄影

牌楼和玺彩画　马炳坚等　摄影

牌楼和玺彩画　马炳坚等　摄影

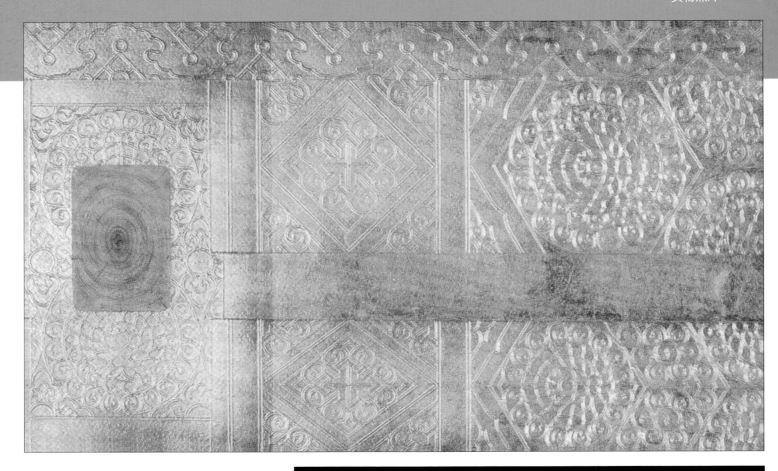

浑金做法旋子彩画 蒋广全供稿（引自于倬云先生主编的《紫禁城宫殿》）

北京八大处六处金龙方心墨线大点金旋子彩画 蒋广全 摄影

实物照片

北京法兴寺内檐墨线大点金旋子彩画　蒋广全　摄影

北京法兴寺内檐墨线大点金旋子彩画（局部）　蒋广全　摄影

实物照片 150

历代帝王庙墨线大点金旋子彩画　蒋广全　摄影

文丞相祠小点金一字方心旋子彩画　高成良　摄影

承德普宁寺西配殿内檐烟琢墨石碾玉旋子彩画　蒋广全　摄影

承德普宁寺钟楼彩画修缮后　蒋广全　摄影

牌楼旋子彩画　马炳坚等　摄影

153 实物照片

一字方心雄黄玉彩画　蒋广全　摄影

椽子、望板彩画　蒋广全　摄影

浑金做法龙雀替　蒋广全　摄影

墙边做法两例　蒋广全　摄影

椽、柁头刷饰　马炳坚等　摄影

椽、柁头彩画　马炳坚等　摄影

椽、柁头彩画　马炳坚等　摄影

掐箍头彩画　马炳坚等　摄影

垂花门包袱式苏画　马炳坚等　摄影

包袱式苏画　马炳坚等　摄影

内檐苏画　马炳坚等　摄影

包袱式苏画　林其浩　摄影

(上)方心式苏画和(下)海墁式苏画　马炳坚等　摄影

方心式苏画　马炳坚等　摄影

实物照片 160

垂头彩画　马炳坚等　摄影

花板彩画　马炳坚等　摄影

清中期方心式苏画(局部)　马炳坚等　摄影

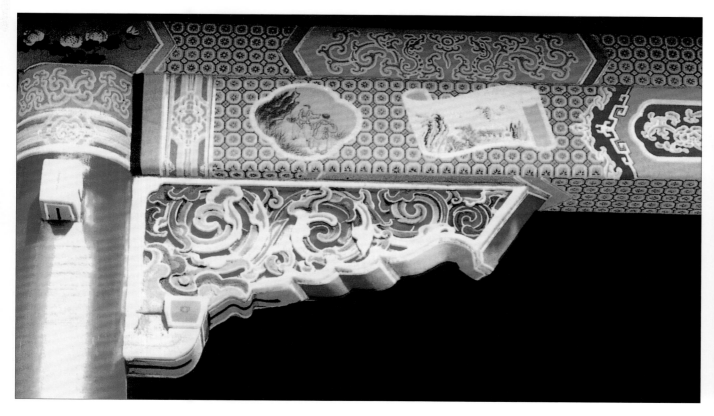

清中期方心式苏画(局部)　马炳坚等　摄影

清中期方心式苏画（局部）　马炳坚等　摄影

清晚期包袱式苏画（局部）　马炳坚等　摄影

清晚期苏画(局部)　马炳坚等　摄影

清晚期苏画(局部)　马炳坚等　摄影

清晚期抱头梁、穿插枋苏画　马炳坚等　摄影

清晚期抱头梁苏画　马炳坚等　摄影

清晚期游廊内檐苏画　马炳坚等　摄影

清晚期游廊内檐苏画（局部）　马炳坚等　摄影

清中期方心式苏画（局部） 马炳坚等 摄影

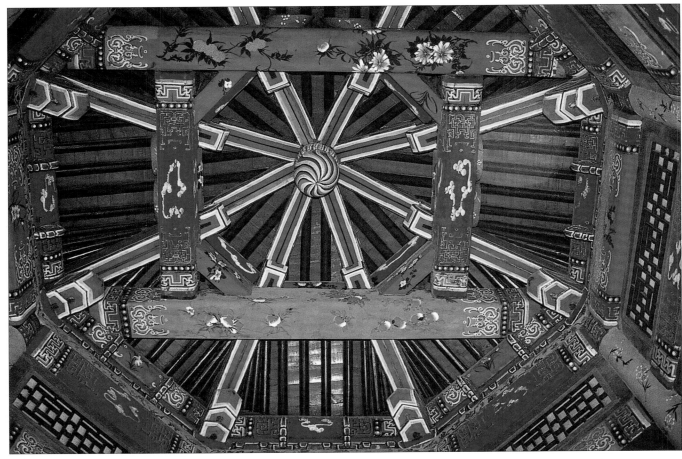

八角亭内檐苏画（局部） 马炳坚等 摄影

苏式彩画包袱（一）线法山水　马炳坚等　摄影

苏式彩画包袱（二）线法人物　马炳坚等　摄影

苏式彩画包袱（三）线法山水　马炳坚等　摄影

苏式彩画包袱（四）线法人物　马炳坚等　摄影

苏式彩画包袱（五）线法人物　马炳坚等　摄影

苏式彩画包袱（六）线法人物　马炳坚等　摄影

苏式彩画包袱（七）线法人物　马炳坚等　摄影

苏式彩画包袱（八）富贵白头　马炳坚等　摄影

苏式彩画包袱(九)花鸟　马炳坚等　摄影

苏式彩画包袱(十)花鸟　马炳坚等　摄影

苏式彩画包袱（十一）松鹤延年　马炳坚等　摄影

苏式彩画包袱（十二）花鸟　马炳坚等　摄影

苏式彩画包袱（十三）百鸟朝凤　马炳坚等　摄影

苏式彩画包袱（十四）海宴河清　马炳坚等　摄影

苏式彩画包袱（十五）富贵满堂　马炳坚等　摄影

苏式彩画包袱（十六）玉堂富贵　马炳坚等　摄影

苏式彩画包袱（十七）三国人物故事　马炳坚等　摄影

苏式彩画包袱（十八）三国人物故事　马炳坚等　摄影

苏式彩画包袱（十九）三国人物故事　马炳坚等　摄影

苏式彩画包袱（二十）三国人物故事　马炳坚等　摄影

苏式彩画包袱（二十二）三国人物故事　马炳坚等　摄影

苏式彩画包袱（二十一）聊斋人物故事　马炳坚等　摄影

苏式彩画包袱（二十三）落墨山水　马炳坚等　摄影

苏式彩画包袱（二十四）包公案人物故事　马炳坚等　摄影

苏式彩画包袱（二十五）落墨山水　马炳坚等　摄影

龙草和玺（1） 王仲杰 蒋广全 冯世怀 张秀芬

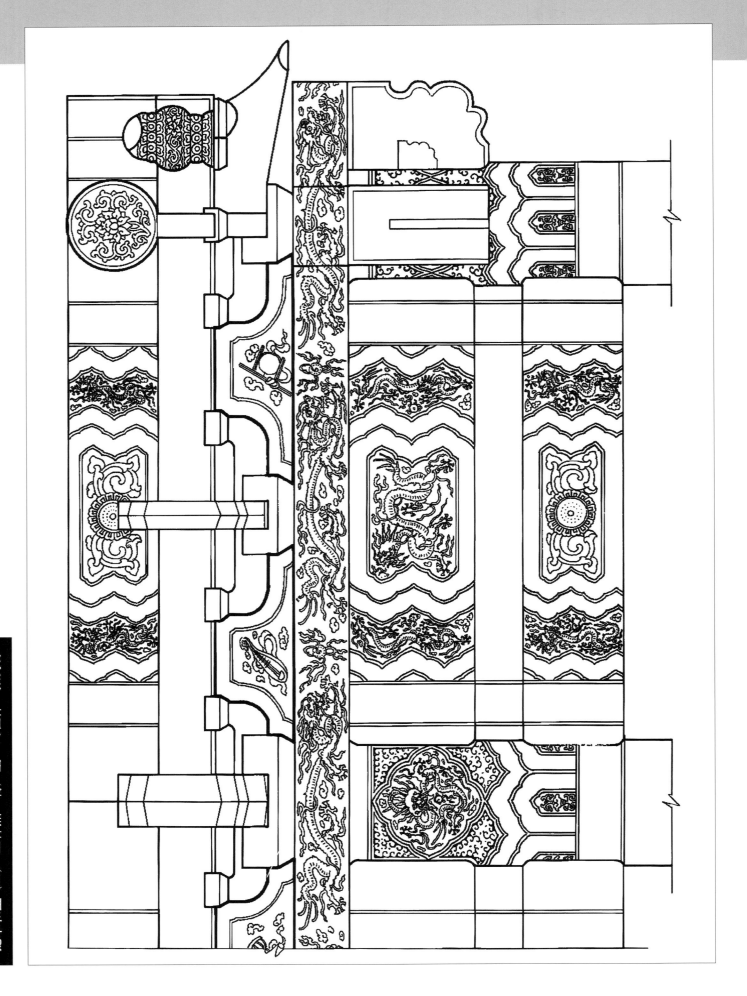

龙草和玺（二） 王仲杰 蒋广全 冯世怀 张秀芬

龙草和玺（三）王仲杰 蒋广全 冯世怀 张秀芬

龙草和玺（四）王仲杰 蒋广全 冯世怀 张秀芬

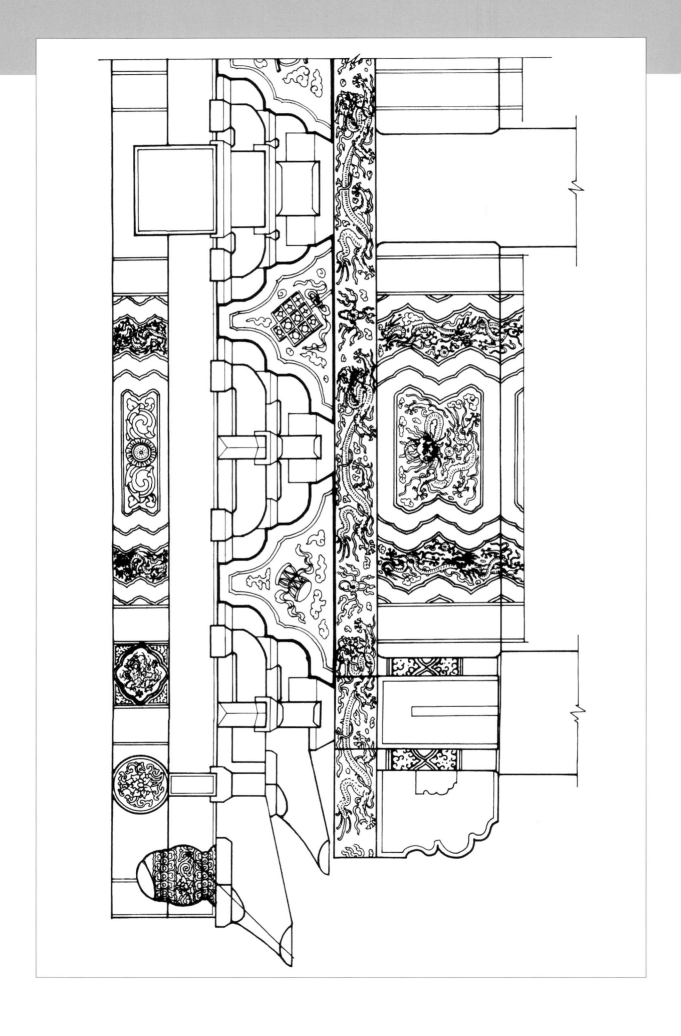

183　墨线图

早期行龙　赵双成　　　　早期升龙　赵双成

早期降龙　赵双成　　　　早期坐龙　赵双成

晚期行龙　赵双成

晚期升龙　赵双成

晚期坐龙　赵双成

晚期降龙　赵双成

夔行龙　赵双成

夔升龙　赵双成

夔坐龙　赵双成

夔降龙　赵双成

墨线图 186

升凤 赵双成

行凤 赵双成

团凤 赵双成

降凤 赵双成

187 墨线图

观头箍头　赵双成

各式箍头　蒋广全

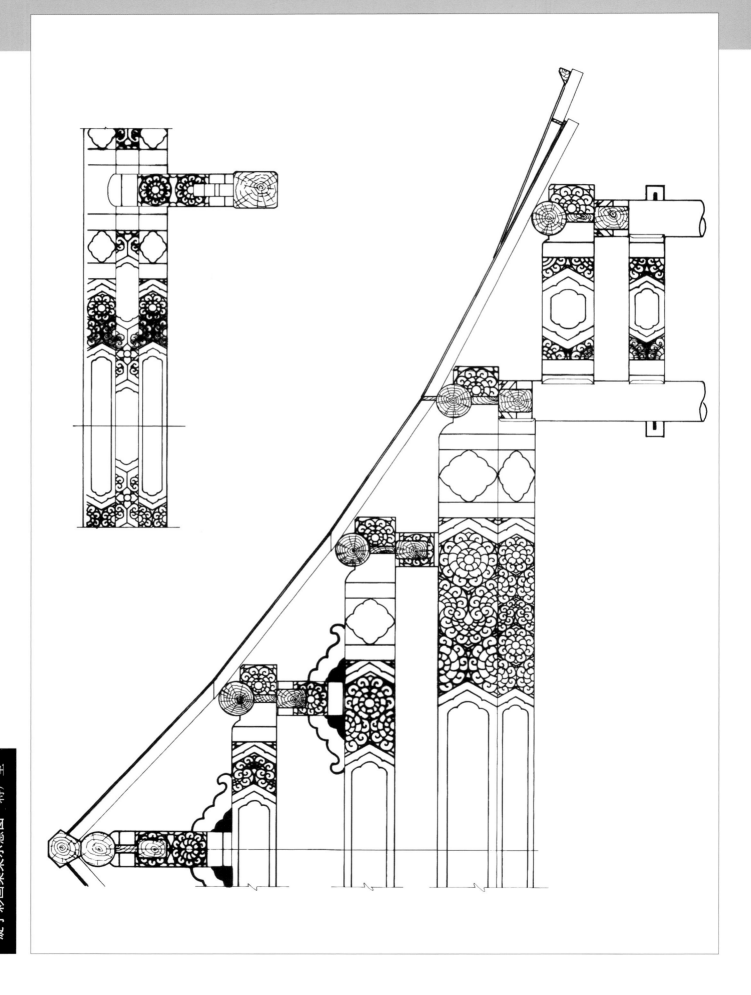

金线大点金旋子加包袱 蒋广全

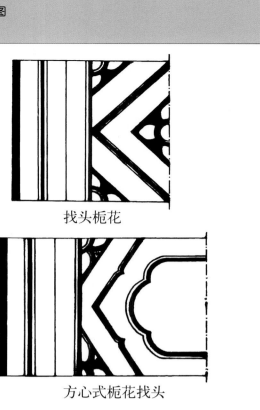

找头栀花

方心式栀花找头

四分之一旋花

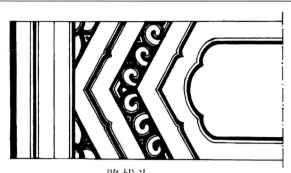

一路找头

金道冠找头

两路旋瓣找头

双金道冠找头

勾丝咬找头

旋子彩画找头　蒋广全

旋子彩画找头　蒋广全

喜相逢找头

一整两破找头

一整两破加一路找头

一整两破加金道冠

平板枋纹饰　赵双成

池子纹饰　蒋广全

异兽　赵双成

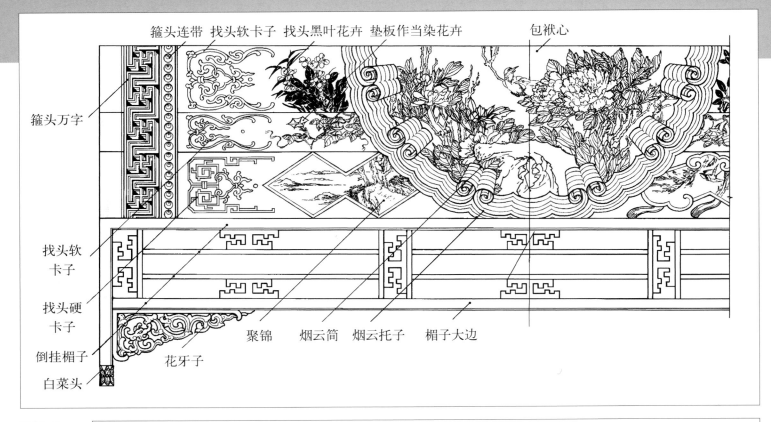

包袱式苏画　蒋广全

海墁式苏画　蒋广全

方心式苏画　蒋广全

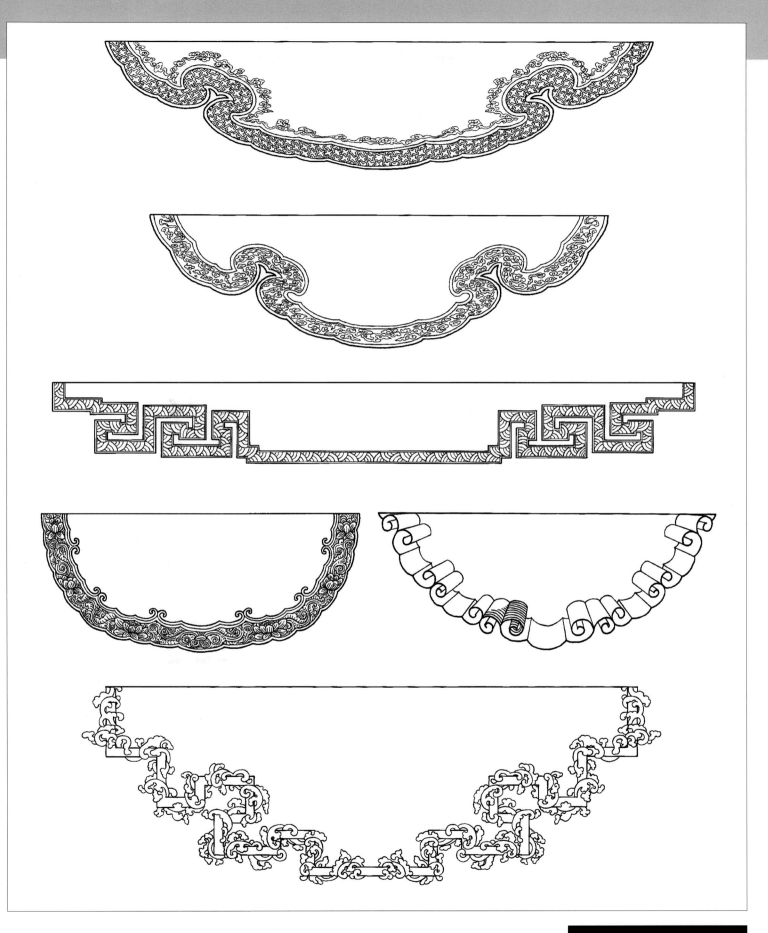

苏式彩画包袱边　蒋广全

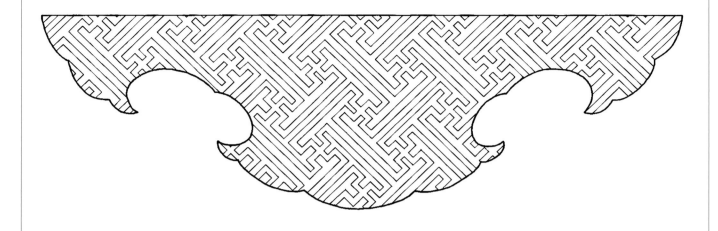

苏式彩画包袱心　蒋广全

包袱心两则　冯世怀

包袱心两则　冯世怀

包袱心两则　冯世怀

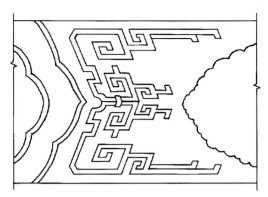

找头片金斜硬卡子

垫板玉做加点金软卡子

垫板玉做软卡子

找头玉做斜硬卡子

找头玉做软卡子

找头玉做软卡子

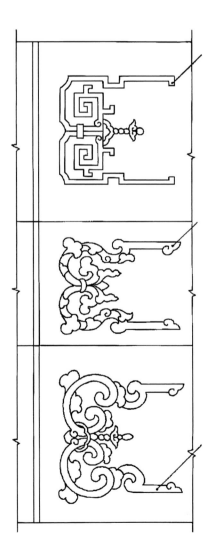

找头软、硬卡子

卡子　蒋广全

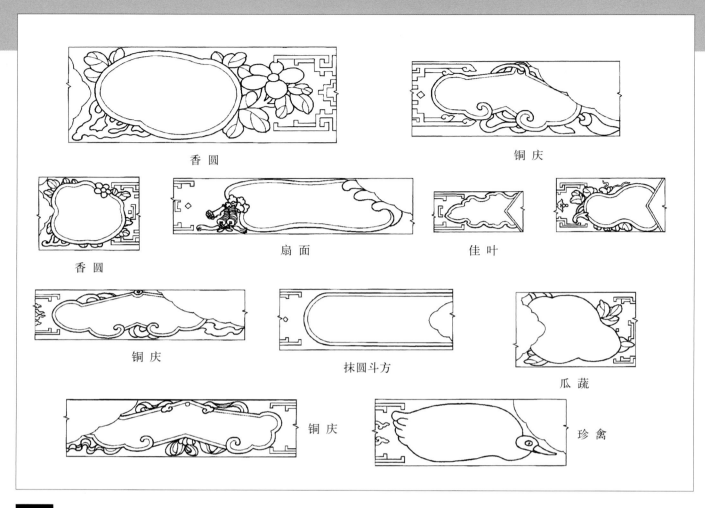

香圆　铜庆　香圆　扇面　佳叶　铜庆　抹圆斗方　瓜蔬　铜庆　珍禽

聚锦壳　蒋广全

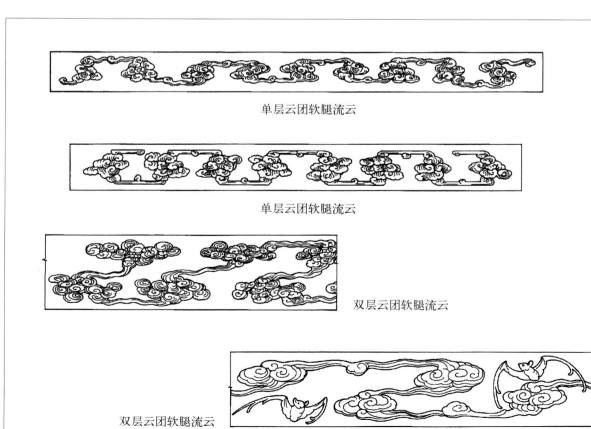

单层云团软腿流云

单层云团软腿流云

双层云团软腿流云

双层云团软腿流云

流云　蒋广全

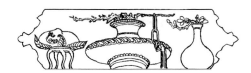
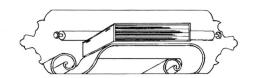
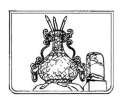

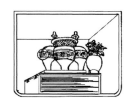

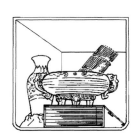
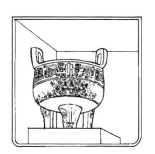
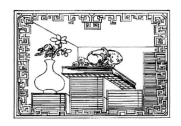

柁头　蒋广全　　　　　　　　　　　　　　　　　　　　　　　　　**柁头帮　赵双成**

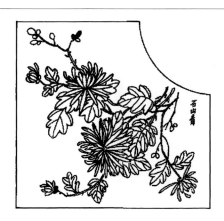
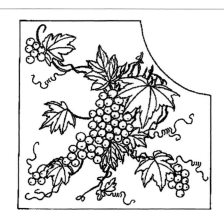
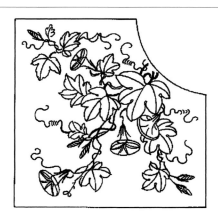

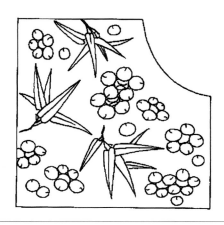
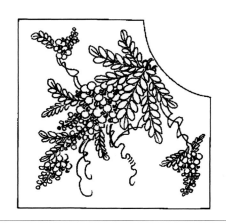

枋底切活 赵双成

坐龙天花　蒋广全

金莲水草天花　蒋广全

双凤天花　蒋广全

双鹤天花　蒋广全

西蕃莲天花　蒋广全

五福（蝠）捧寿天花　蒋广全

宝珠吉祥草彩画　王仲杰

斑竹纹海墁彩画　蒋广全

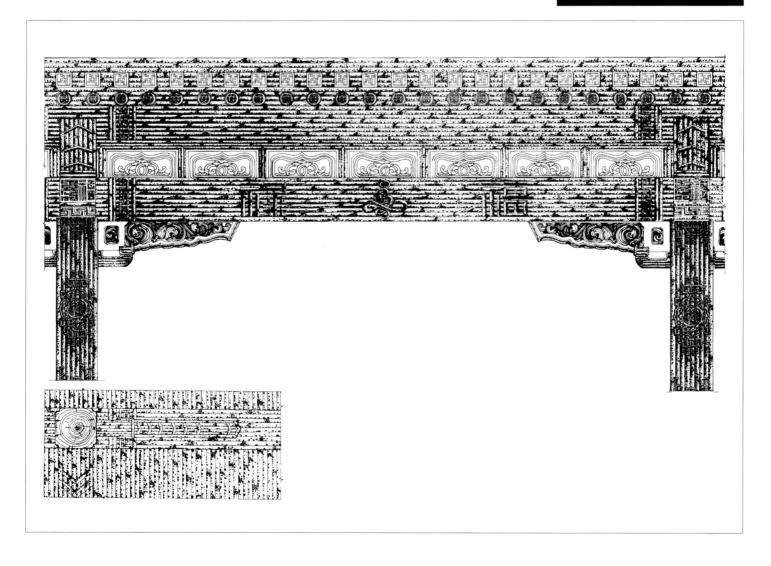

新式柱纹三则　赵双成

新式彩画灯花四则　赵双成